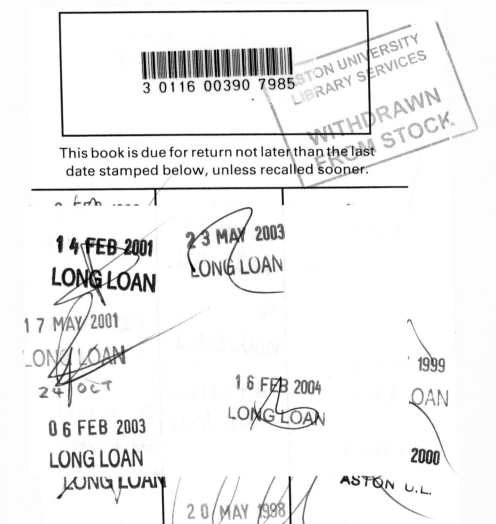

SUCCESSFUL MANAGEMENT FOR SAFETY

Organizing Committee

B. Crossland
A. Canning
D. Eves
B. Harding
J. Lakey

P. Makin
B. Mansell
K. Millard
S. Nicholls
B. Scrase

Associated bodies

Hazards Forum
British Safety Council
Health and Safety Executive
Health and Safety Group of the SCI (Society of Chemical Industry)
Institute of Management
Institute of Marine Engineers
Institution of Civil Engineers
Institution of Electrical Engineers
Institution of Nuclear Engineers
Institution of Occupational Safety and Health
Royal Society for the Prevention of Accidents

Cover illustration courtesy of Ford Motor Company

SUCCESSFUL MANAGEMENT FOR SAFETY

Papers presented at a Meeting organized by the Engineering Manufacturing Industries Division of the Institution of Mechanical Engineers in association with the Hazards Forum, and held at The Institution of Mechanical Engineers on 12–13 October 1993.

Published by
Mechanical Engineering Publications Limited for
The Institution of Mechanical Engineers
LONDON

© The Institution of Mechanical Engineers 1993

This publication is copyright under the Berne Convention and the International Copyright Convention. All rights reserved. Apart from any fair dealing for the purpose of private study, research, criticism or review, as permitted under the Copyright, Designs, and Patents Act, 1988, no part of this publication may be reproduced, stored in a retrieval system, or transmitted in any form or by any means, without prior permission of the copyright owners. Reprographic reproduction is permitted only in accordance with the terms and licences issued by the Copyright licensing Agency, 90 Tottenham Court Road, London W19 9HE. Unlicensed multiple copying of the contents of the publication without permission is illegal. Inquiries should be addressed to: The Managing Editor, Mechanical Engineering Publications Limited, Northgate Avenue, Bury St Edmunds, Suffolk IP32 6BW, England.

Authorization to photocopy items for personal or internal use, or the internal or personal use of specific clients, is granted by the Institution of Mechanical Engineers for libraries and other users registered with the Copyright Clearance Center (CCC), provided that the base fee of $3.00 per paper plus $0.05 per page is paid direct to CCC, 21 Congress Street, Salem, Ma 01970, USA. This authorization does not extend to other kinds of copying, such as copying for general distribution, for advertising or promotional purposes, for creating new collective works, or for resale. No copying fees are payable for papers published prior to 1978.
0957-6509/93 $3.00 + 0.05

The publishers are not responsible for any statement made in this publication. Data, discussion and conclusions developed by authors are for information only and are not intended for use without independent substantiating investigation on the part of potential users. Opinions expressed are those of the authors (or contributors to discussion) and are not necessarily those of the Institution of Mechanical Engineers or its publishers.

ISBN 0 85298 858 3

A CIP catalogue for this book is available from
the British Library.

Printed by Moreton Hall Press Ltd,
Bury St. Edmunds, Suffolk

CONTENTS

The impetus from legislation — D. Eves — 1

The Community's contribution as regards safety and health in the context of completion of the internal market – the social point of view — A. Pangalos — 7

Assessment of a company for safety and minimization of losses — I. Neil — 19

Quality risk and safety — F. Warner — 29

The role of human factors and safety culture in safety management — R. T. Booth and T. R. Lee — 43

Managing professionally — A. Osborne and D. Brown — 53

Company strategy for the management of safety — F. T. Duggan — 65

New technology – the implications for management — V. L. Mayatt — 71

Safety and new technology in aviation – providing the regulatory framework — D. A. Whittle — 75

Management responsibility for the safety of software — D. W. Newman — 81

Safety management of the Sizewell B project — B. V. George — 85

Legal consequences of accidents – managing the product liability issue — A. J. Hobkinson — 95

The management of safety and emergency planning — G. D. Kenney — 105

Crisis management and a corporate response organization — M. W. Howard — 111

If you live in England and Wales you are a user of nuclear electricity.

One fifth of the electricity we all need is generated by nuclear power.

At Knutsford in Cheshire, Nuclear Electric's **PWR Project Group** is the driving force behind the construction of Britain's newest nuclear power station, **Sizewell B** in Suffolk.

Main control room, Sizewell B.

From construction through to generation, Nuclear Electric staff and their contractors are dedicated to ensuring that Sizewell B meets the highest safety standards.

The power from Sizewell B will help to maintain the quality of life not only for today but for future generations – **Safely**

If you would like to know more about Nuclear Electric, please contact Liz Griffiths, Public Relations Officer on 0565 682604.

Generating energy *for* generations

Nuclear Electric

YOU CAN'T TAKE CHANCES IF YOU'RE AIMING FOR THE TOP

Mountaineering is dangerous enough as it is, without taking unnecessary chances. Professional mountaineers cut risks to a minimum by making meticulous plans and following carefully-developed safety practices.

There's a moral here for the business world. Every commercial undertaking has inherent risks, but successful companies make the elimination of avoidable losses a top priority. And that's where we can help.

Our experienced consultants can work with you to develop the safety management strategies which will reduce risk to a minimum and ensure that you meet European and national legislative requirements. And, naturally, we provide the training and auditing you'll need to support your initiatives.

By creating a safety-oriented culture based on sound Quality principles you will be doing everything possible to protect your employees, yourself and, of course, your business. So why not talk to us and find out how we can help you aim high, without taking high risks?

Certificate No. FS 22318
BS 5750 : Part 1. ISO 9001

Management Consultants

Gilbert Associates (Europe) Limited, Fraser House, 15 London Road, Twickenham TW1 3ST.
Telephone: 081 891 4383, Facsimile: 081 891 5885

ASTON UNIVERSITY, Health & Safety Unit,
Department of Mechanical & Electrical Engineering.
Aston Triangle, Birmingham, B4 7ET.
Tel: 021-359-3611 Ext 5194 or 4332.

ASTON UNIVERSITY

In January 1994 we are offering for the first time an MSc/ Postgraduate Diploma in Risk Management and Safety Technology (full or part-time). A major driving force behind this new programme is the Engineering Council's Code of Professional Practice, *"Engineers and Risk Issues"*. Practising engineers who are seeking a career move into safety management may be particularly interested. The course fees are comparable with other vocational Masters programmes. Write or telephone Dr Mark Cooper for further details.

MECHANICAL ENGINEERING PUBLICATIONS

ENGINEERING SYSTEM SAFETY

By G J Terry C Eng FIMechE

Many aspects of safety are covered by regulations and codes of practice, but ere is no clear guidance, or where additional requirements are appropriate. It is in eer is faced with many conflicting demands, that the needs of safety can so easily

It is here that **Engineering System Safety** provides invaluable guidance t signer, and the engineering manager, enabling them to make realistic and well-informed decisions about ty of any engineering system.

Engineering System Safety provides excellent advice on how to justify decisions about safety, and how to assemble arguments and data to support claims for the safety of a system, from small single components through to major industrial processes.

0 85298 781 1/210 x 148mm/softcover/154 pages/November 1991
£34.00 inclusive of *free* delivery to a UK address. Overseas customers please add 10% for delivery.

Telephone orders using Visa/Mastercard welcome.
Please ring Sales Department Direct line 0284 724384

Orders and enquiries to: Sales Department, Mechanical Engineering Publications Limited,
Northgate Avenue, Bury St. Edmunds, Suffolk IP32 6BW, England.
Tel: (0284) 763277 Fax: (0284) 704006 Telex: 817376

The impetus from legislation

D EVES, BA, CB, FIOSH,
Health and Safety Executive, London, UK

Introduction

HSE believes that effective management of health and safety in the workplace is a fundamental task which employers cannot afford to ignore. This paper explains:
- why health and safety management is important;
- how the role of management has been reflected in health and safety legislation, in particular the Management of Health and Safety at Work Regulations 1992;
- that good management of health and safety cannot be achieved by legislation alone.

The Importance Of Management Of Health And Safety

HSE's inspectors still meet employers who think that all they need do to manage health and safety in their workplace is to appoint a good safety officer and then leave all health and safety matters to that person. Or they may decide to engage a consultant to relieve them of much of the thinking. But although employers are required by law to have competent assistance in this area, health and safety is too important to be left to one or two people in an organisation. It is not simply a technical matter that can be pushed to one side, to be considered more widely only in the event of a serious incident. In fact, most accidents - and their costs - can be avoided if the management systems are good enough. Successful companies understand this, and align health and safety closely with other business goals..
Major accidents often cost dearly, in both human and financial terms. The Piper Alpha disaster involved the loss of 167 lives and is estimated to have cost over £2 billion. BP have estimated that the Grangemouth refinery explosion, in which one person died, cost £100 million, about half being attributed to business interruption.
Although disasters on this scale are relatively infrequent in the UK, every day people are either killed or injured at work or their health is affected. We estimate that 31,000,000 working days were lost through accidents or illhealth at work in 1991/92, and that the cost to the nation was around 1-2% of GDP. In 1991/92 466 fatal injuries were reported to HSE. Sometimes it is a matter of chance as to whether an accident affects health and safety. Many accidents do not involve loss of life or injury or ill-health, but they cost money through damage to plant and equipment or by affecting business efficiency in some other way.
Studies have shown us that few employers realise either the true cost of this financial haemorrhage or the extent to which accidents can be avoided. When these two points are put together, the conclusion is inescapable: employers cannot afford to ignore the potential savings that come from an effective approach to managing health and safety.

HSE Case Studies

Just how much accidents cost has been demonstrated by case studies undertaken by HSE in conjunction with five employers in different parts of industry and published in "The Costs of Accidents At Work" (1). Briefly, the work was carried out in the construction, food, transport and oil industries and a hospital. Each organisation's health and safety performance was at least average for their industry. The studies ran for three months each, using an agreed methodology to quantify fully the costs associated with the accidents recorded. An accident was defined as any unplanned event that resulted in injury or ill health to people, or damage or loss to property, plant, materials or the environment or a loss of business opportunity. All personal injuries were recorded along with other accidents that were regarded by the organisations to be preventable and to cost over a given amount. Figure 1 shows that the losses identified were substantial in every case.

FIGURE 1

SUMMARY OF LOSSES IDENTIFIED[1]

	TOTAL LOSS (000's)	ANNUALISED LOSS (000's)	REPRESENTING
1 Construction Site	£245	£700[2]	8.5% of tender price
2 Creamery	£244	£975	1.4% of operating costs
3 Transport Company	£49	£196	1.8% of operating costs 37% of profits
4 Oil Platform	£941	£3,764	14.2% of potential output
5 Hospital	£99	£397	5% of annual running costs

(1) Figures quoted are actual at time of study: no adjustment has been made for inflation. Study[1] lasted 18 weeks; studies 2 - 5 13 weeks each.
(2) Represents total length of contract (54 weeks).

The ratio of insured to uninsured costs incurred by the organisations were also compared. These are portrayed in Figure 2 as an iceberg: visible costs above the water-line represent different insurance premia paid; the hidden, uninsured costs below the water-line were between 8 and 36 times what was being paid in insurance premia.

Figure 2

Accident Iceberg - The Hidden Cost Of Accidents

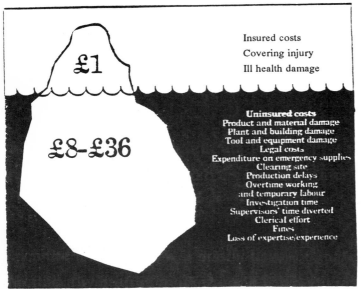

The costs identified by the study surprised HSE just as much as the participating organisations. Bearing in mind that the accidents in question could have been prevented, cost-conscious managers must take seriously the potential gains that accrue from an effective health and safety system.

Health and Safety Legislation

The role of management in health and safety is central to the Health and Safety at Work etc Act 1974. The Act is based on the principle that the primary responsibility for ensuring health and safety in the workplace lies with those who create the risks. This gives rise to the general duty "of every employer to ensure, so far as is reasonably practicable, the health, safety and welfare at work of all his employees" (and others who may also be affected) (2).

We have used the opportunity to build on rather than change our existing legislative system for health and safety. Until this year, some of the actions necessary for employers to discharge their duties under the Act were implicit in the law. Now, the Management of Health and Safety at Work Regulations 1992 (the Management Regulations) (3) spell out these duties more explicitly. These regulations help implement the European Community Framework Directive on health and safety. They are part of the "six-pack" of regulations, all of which came into force on 1 January 1993 to implement EC directives designed to improve the general levels of health and safety across the European Community.

The Management Regulations focus on the key components of managing health and safety. For example, they require employers to:

- assess the risks associated with their work activity, so that the necessary preventive and protective measures can be identified and put in place:

- make arrangements to cover health and safety, ie effective planning, organisation, control, monitoring and review;

- provide health and safety surveillance where necessary;

- appoint competent people (from inside or outside the organisation) to help devise the measures that need to be taken;

- set up emergency procedures;

- provide employees with the information and training they need to do their jobs safely;

- co-operate with other employers as necessary.

The Function of Legislation

The Management Regulations were needed to fulfil obligations arising from our membership of the EC. But it is worth examining in more general terms some of the considerations which lie behind whether to use legislation to address a particular issue.

Although legislation has the potential to change both behaviour and attitudes, by its very nature it must be directed primarily at the former rather than the latter.

Laws can be passed and enforced governing how people must **act**; however, laws governing how people **think** are a different prospect altogether.

Of course, the very act of legislating is a stimulus which can change people's attitudes. The public debate and other publicity that usually surrounds a change in the law can help people to understand why legislation is needed. An appreciation of the issues will affect the public's attitude towards a new law. The wearing of car seat belts is a good example.

It is not always easy however to establish the extent to which a change in either behaviour or attitude has been due to the law or to a variety of other factors. Factors other than legislation can both change attitudes and behaviour, eg as the result of publicity campaigns by pressure groups or government, through the increased use by the courts of existing penalties, or by subtle, almost indefinable changes in the way that society views a particular action.

HSE would much prefer employers and workers to be motivated to observe good health and safety practice (be it contained in guidance or law) because of realisation that this will actually help them carry out their activities in a way that minimises the risk to themselves, to their business and to others. However, we sometimes have to settle for legislation which changes behaviour, even if we are less sure that attitudes have been changed. For example, we would have preferred workers to wear head protection on construction sites because they realised that this is an effective measure of reducing risk of injury. I am bound to say that in the absence of regulations hard hats were rarely worn. However, since the making of the Construction (Head Protection) Regulations 1989, there has been a dramatic improvement. In this sort of example changes in behaviour can be measured with relative ease. We are convinced that these Regulations had more impact in changing behaviour than previous publicity campaigns and appeals to common sense.

However, there are drawbacks to an over-reliance on legislating every time a problem is identified. Vital considerations for government are the potential costs, benefits and effectiveness of any new law. Legislation has to be carefully framed so that it fulfils its intentions and can be properly enforced; an ill-conceived law can actually be counter-productive. Moreover, the sheer volume of regulation can be daunting to employers, particularly small businesses. Also, each new law represents a potential curtailment of individual choice. Governments therefore need to think very carefully before legislating; they should be convinced that their objectives cannot be attained as effectively by other means.

Setting out to change behaviour through legislation alone is not always going to be sufficient. If the legislation is really going to achieve its objective, a change of attitude is essential. I suggest that the management of health and safety is an obvious case. For example, without firm commitment from management, risk assessments and the actions that should flow from them can become mechanistic and bureaucratic exercises, carried out simply to demonstrate compliance with the letter of the law but with little thought to the reason for the assessment. To derive the maximum benefit from the Management Regulations both behaviour and attitude have to be right, and this becomes manifest in the creation of a positive health and safety culture within the organisation. This cannot be achieved by legislation alone.

Developing a Health and Safety Culture

The role of management is crucial to the development of a positive health and safety culture. Our studies have shown that only if management are genuinely committed to health and safety are there likely to be effective policies and procedures throughout the rest of the organisation.

Investigation and research shows that accidents and ill health are rarely inevitable random events. They generally arise from failure in control and have multiple causes. Therefore, the most effective method of improving health and safety performance is by applying good management techniques.

These techniques need not be complex. HSE has produced guidance to help managers, in particular a booklet called "Successful Health & Safety Management" (4) and an accompanying leaflet (5). Both identify five key steps:

1) **set the policy**: aim to minimise and control all accidental loss; record the policy and implementation arrangements so that what has been done is clear for staff and others;

2) **organise the staff**: managers need the commitment and participation of staff; staff need to be competent; responsibilities must be clearly allocated; staff should be consulted and involved in solving problems; they should be informed about risks and measures taken to control them;

3) **plan and set standards**: planning is the key to ensuring that health and safety

arrangements really work; identify hazards and assess risks; then decide what precautions are needed and put them in place;

4) **measure performance**: managers need to know if measures are working; active monitoring (before things go wrong) establishes that procedures are in place and are working; reactive monitoring (after things go wrong) involves learning from mistakes;

5) **learn from experience**: review performance regularly using information from monitoring and elsewhere (eg, from independent audit).

There is nothing novel about these steps. They simply set down for health and safety what is good management practice in other areas of work. Furthermore, good managers will often be able to incorporate health and safety issues into more general management activities (such as planning and monitoring) rather than considering them in isolation. For example, a vital task of management in a manufacturing industry is to ensure that there are adequate systems to control production costs. As HSE's studies have shown, health and safety has an important part to play in keeping costs down by minimising accidents at work; health and safety should therefore be a central feature of a product control system.

Proper attention to health and safety does not have to mean introducing complex or costly procedures. Good managers will ensure that health and safety measures are proportionate to the risks involved. This approach means thinking about the problems and identifying appropriate solutions. This need not be difficult, expensive or time consuming, but it is done most effectively within the context of an established health and safety culture.

The Role of Engineers

Many employers and managers still have not taken these messages on board. What more can be done? In developing a safety culture within any manufacturing organisation, those responsible for engineering safety have crucial roles to play. Engineers of all disciplines, guided by their Institutions, have played the principal role in developing and setting the technical standards that are the foundation of engineering integrity and fitness for purpose. Engineers have also played a leading role in the engineering of safety (ie, the management and control of all the systems that contribute to a safe and healthy working environment).

The professional Institutions, through control of the accreditation of university degree courses, the management of professional development to chartered status and monitoring of continuing professional development can influence the inculcation of positive attitudes towards safety in those who will have statutory responsibilities for health and safety in their companies. HSE regards it as vital that the professional Institutions continue to play their part in achieving an improved safety culture throughout the ranks of managers in industry.

It is therefore very encouraging that the Institutions, through the Engineering Council and the Hazard Forum, have been emphasising the importance of risk assessment. A knowledge of risk assessment is a fundamental technique and an essential attribute for all those with responsibility for making technical decisions that impact on health and safety.

This is an area in which HSE has been prominent in recent years and the incorporation of a requirement for risk assessment into the Management Regulations is welcome recognition of the importance of this subject to engineers and managers alike.

Conclusion

The Management Regulations are already changing the behaviour of many managers but will not be fully effective without a corresponding change in attitude. The law tells managers what must be done; HSE's guidance will help them to do it. But managers also need to understand why they should devote attention to health and safety management.

Research and experience demonstrate the importance of managing health and safety. Good management is essential in spreading a health and safety culture throughout an organisation. This in turn reduces the losses caused by accidents and illhealth - as well as the human suffering that accompanies death or serious injury.

HSE aims to convey these messages and help influence attitudes in a variety of ways - by legislation, by publicity and by opportunities such as this. Legislation certainly provides an impetus. It emphasises the significance which government attaches to the issue; it enables HSE to use inspection and enforcement powers to guide (and sometimes compel) employers to take the necessary actions. However, legislation must be combined with publicity, education and persuasion if it is to have the desired effect throughout the whole of industry.

References:

(1)   The Costs of Accidents at Work
        published by HMSO - ISBN 0
        11 886374 6

(2)   Health and Safety at Work etc Act 1974
        published by HMSO -
        ISBN 0 10 543774 3

(3)   Management of Health and Safety at Work
        Regulations 1992 and Approved Code of
        Practice published by HMSO - ISBN 0
11   886330 4

(4)   Successful Health & Safety Management
        published by HMSO -
        ISBN 0 11 885988 9

(5)   Five Steps to Successful Health and Safety
        Management free leaflet available from
        HSE

The Community's contribution as regards safety and health in the context of completion of the internal market – the social point of view

A PANGALOS
DGV, Commission of the European Communities, Luxembourg

I have to thank the organizers for giving me the opportunity to present the Community activities in relation to the social dimension of the internal market, especially in the field of health and safety.

In my speech, I will briefly describe the situation before 1985, the year in which the idea of the internal market got a fresh impulse with the "White Paper". Following this, I will outline the changes brought about by the Single Act and the new legislative possibilities it offers; then, the Commission's action programme on health and safety, the Social Charter and the directives adopted up to now. After a short review of the main Maastricht decisions I will conclude with a look at future activity.

Since the early years of the Community, safety was treated under two different aspects :

- in a programme for the elimination of technical barriers to trade concerning the intrinsic qualities and risks of products, and

- in an action programme in the field of health and safety, focussing on the use of products by workers.

However it appeared that the process of decision-making, namely unanimity, hindered the Community in developing rapidly into a Common market. In 1985, the Commission's White Paper on completion of the internal market listed the efforts to be made. The "new approach" was introduced, reducing the content and size of legislative texts to a reasonable but abstract level, and relying upon standardization for the elaboration of the technical details. In 1987, the ratification of the European Single Act with the important Articles 100A (on the free movement of goods) and 118A (safety and health protection at work) added this new spirit to the Treaty of Rome. Article 118A is of particular interest to the directorate of "health and safety", as nearly all our legislative work is based on this article.

Let me recall the meaning of Internal Market.

It is defined (Article 8A EEC) as comprising "an area without frontiers in which the free movement of goods, persons, services and capital is ensured in accordance with the provisions of this Treaty".

The free circulation of products is to be achieved in particular through the elimination of technical barriers to trade, and in many cases it is in safety requirements that such barriers lie. Many products which ought to be able to circulate freely are items of work equipment or are in some way connected with workplaces.

The new tools available for completion of the internal market are the new approach and Article 100A EEC.

The "new approach to technical harmonization and standards" (Council Resolution 85/C 136/01 of 7.5.1985 OJEC No C 136 of 4.6.1985, p.1) provides that, for completion of the internal market, directives in this field shall contain (only) essential safety requirements and that the technical details shall be left to European standardization organizations.

The legal base is provided by Article 100A of the EEC Treaty (added to the Treaty pursuant to the Single Act of 1987). Some extracts from it:

"1. .... The Council shall, acting by a qualified majority on a proposal from the Commission in cooperation with the European Parliament and the Economic and Social Committee, adopt the measures for the approximation of the provisions laid down by law, regulation or administrative action in Member States which have as their object the establishment and functioning of the internal market.

2. ...

3. The Commission, in its proposals ... concerning health, safety, environmental protection and consumer protection, will take as a base a high level of protection.

4. If, after the adoption of a harmonization measure by the Council acting by a qualified majority, a Member State deems it necessary to apply national provisions on grounds of major needs referred to in Article 36, or relating to protection of the environment or the working environment, it shall notify the Commission of these provisions.

5. The harmonization measures referred to above shall, in appropriate cases, include a safeguard clause authorizing the Member State to take, for one or more of the non-economic reasons referred to in Article 36, provisional measures subject to a Community control procedure."

To sum up: From the social point of view there are two essential points:
- the high level of protection
- the safeguard clause.

As regards standards, they remain voluntary; however, a product's compliance to them provides it with a "presumption of conformity" to the essential safety requirements. The responsibility of each Member State remains unchanged, as it may, through the safeguard clause, oppose either a product or a standard itself. As this is linked to a very stringent procedure of conciliation, the general feeling is that it will not be involved abusively.

The need to give a social dimension to the completion of the internal market was recognized swiftly at Community level by:

- the Commission: in its Communication of October 1987 (88/C 28/02-OJ C 28 of 3.2.1988) on its programme concerning safety, hygiene and health at work.
- the Council: in its Resolution of 21.12.1987 (88/C 28/01-OJ C 28 of 3.2.1988), in which the Council welcomes and supports the Commission programme.

In this Resolution, the Council stresses e.g. the need "to place equal emphasis on achieving the economic and social objectives of the completion of the internal market".

The Commission has reaffirmed its position on this subject with the proposal of a Social Charter (Social Europe No 1/90, Commission of the European Communities) which contains the following chapters:

- Right to freedom of movement
- Employment and remuneration
- Improvement of living and working conditions
- Right to social protection
- Right to freedom of association and collective bargaining
- Right to vocational training
- Right of men and women to equal treatment

- Right to health protection and safety at the workplace
- Right of workers to information, consultation and participation
- Protection of children and adolescents
- Elderly persons
- Disabled persons

The Social Charter which has been adopted by all but one of the Member States is a solemn declaration and does not contain any binding prescriptions. The Commission has presented an accompanying programme dealing with the concrete applications of these principles for each chapter (Commission of the European Communities: Social Europe N 1/90).

Furthermore, the social dimension of the internal market is made up of other items. On the one hand, the Commission endeavours to provide the directives on free movement of goods with a social component by formulating the essential safety requirements in order to achieve a high safety level. The Commission provides also for an appropriate participation of the social partners in standardization work.

On the other hand the concept of Social Dialogue at European level was added through Article 118B of the Treaty, completing the existing provisions at national and company level.

Article 118B states :

"The Commission shall endeavour to develop the dialogue between management and labour at European level ...."

The Commission goes even a step further by involving employers and workers in the drafting of directives based on Articles 100 A (essential safety requirements) and 118 A (minimum requirements), and is endeavouring to do likewise in the drafting of European standards.

Without excluding possible ad hoc consultations, the main forum in which employers and workers can exercise their role at Community level is the Advisory Committee on Safety, Hygiene and Health Protection at Work (set up by Council Decision in 1974).

Since all proposals for directives on the safety and health of workers drafted on the basis of Articles 118 A and 100 A (as well as the concomitant standardization mandates) are transmitted to this Committee for its opinion, it is clear that the activities of the Committee have developed considerably.

The Maastricht summit has provided now for another major axis for the involvement of social partners. Unfortunately it was only possible to find an agreement between 11 Member States, so that the new provisions do not form part of the new Treaty but are only attached to a protocol.

This agreement foresees consultations of the social partners on the orientation to give to a planned action and later, its contents. At this moment the social partners may chose to regulate the subject on the basis of a collective agreement.

Even if the subject is treated by the Council there is still a possibility that the transposition into a Member States' legislation is superseded by a collective agreement at the national level.

Let us now come back to some explanations about the legal basis of our activities, Article 118A EEC and its specifics as opposed to Article 100A. The complete text reads as follows:

"1. Member States shall pay particular attention to encouraging improvements, especially in the working environment, as regards the health and safety of workers, and shall set as their objective the harmonization of conditions in this area, while maintaining the improvements made.

2. In order to help achieve the objective laid down in the first paragraph, the Council, acting by a qualified majority on a proposal from the Commission, in cooperation with the European Parliament and after consulting the Economic and Social Committee, shall adopt, by means of directives, minimum requirements for gradual implementation, having regard to the conditions and technical rules obtaining in each of the Member States.

Such directives shall avoid imposing administrative, financial and legal constraints in a way which would hold back the creation and development of small and medium-sized undertakings.

3. The provisions adopted pursuant to this Article shall not prevent any Member State from maintaining or introducing more stringent measures for the protection of working conditions compatible with this Treaty."

To sum up, Article 118A foresees directives to be acted by a qualified majority through the cooperation procedure with the European Parliament. Minimum requirements shall be fixed, but each Member State is free to ask for more. This means that in the near future, the situation in this field will not lead to a complete harmonization and that national deviations will remain. Furthermore, this Article 118A requires special treatment of small and medium enterprises, be it through different deadlines of application or less stringent administrative dispositions.

In opposition to Article 100A, the new approach will not be applicable insofar as harmonized European standards are not foreseen to give the necessary technical details to the provisions of directives under Article 118A EEC. This will be done by the national authorities. Only very specific items such as measurement methods could be committed to CEN/CENELEC.

Let us now turn to the third action programme and its activities which are directly geared towards the internal market.

The main themes are:

A) safety and ergonomics at work

    1. cooperation in establishing the essential safety requirements and European standards
    2. harmonization of safety at work and application of ergonomic principles
    3. safety in high-risk sectors

B) occupational health and hygiene

C) information

D) training

E) small and medium-size undertakings

F) the social dialogue.

A number of directives have been proposed and adopted on the basis of article 118A, referring to the third action programme.

The first one, the so-called "framework directive" introduces measures to encourage improvements in the safety and health of workers at the workplace and gives the general framework for the other directives which are based on it : the Council Directive of 12 June 1989 on the introduction of measures to encourage improvements in the safety and health of workers at work (89/391/EEC, OJEC No L 183, 29.6.1989, p.1).

Up to now the following individual directives have been adopted by the Council :

- Council Directive of 30 November 1989 concerning the minimum safety and health requirements for the workplace, (89/654/EEC, OJEC No L 393, 30.12.1989, p. 1

- Council Directive of 30 November 1989 concerning the minimum safety and health requirements for the use of work equipment by workers at work (89/655/EEC, OJEC No L 393, 30.12.1989, p. 13)

- Council Directive of 30 November 1989 on the minimum health and safety requirements for the use by workers of personal protective equipment at the workplace (89/656/EEC, OJEC No L 393, 30.12.1989, p. 18)

- Council Directive of 29 May 1990 on the minimum health and safety requirements for the manual handling of loads where there is a risk particularly of back injury to workers (90/269/EEC, OJEC No L 156, 21.6.1990, p. 9)

- Council Directive of 29 May 1990 on the minimum safety and health requirements for work with display screen equipment (90/270/EEC, OJEC No L 156, 21.6.1990, p. 14)

- Council Directive of 28 June 1990 on the protection of workers from the risks related to exposure to carcinogens at work (90/394/EEC, OJEC No L 196, 26.7.1990, p.1)

- Council Directive of 26 November 1990 on the protection of workers from risks related to exposure to biological agents at work (90/679/EEC, OJEC No L 374 of 31 December 1990, p.1)

- Council Directive of 25 June 1991 amending Directive 83/477/EEC on the protection of workers from the risks related to exposure to asbestos at work (91/382/EEC, OJEC No L 206 of 29.7.91, p.16)

- Council Directive of 31 March 1992 on the minimum health and safety requirements for improved medical treatment on board vessels (92/29/EEC, OJEC N L113 of 30.4.92, p. 19)

The Commission has also adopted :

- Council Directive of 24 June 1992 on the implementation of minimum safety and health requirements at temporary or mobile work sites (92/57/EEC, OJEC N L245 of 26.08.92, p. 6)

- Council Directive of 24 June 1992 concerning the minimum requirements for the provision of safety/or and health signs at work (92/58/EEC, OJEC N L245 of 26.08.92, p. 23)

- Council Directive of 19 October 1992 on the introduction of measures to encourage improvements in the safety and health at work of pregnant workers and workers who have recently given birth or are breastfeeding (92/85/ECC, OJEC N L348 of 28.11.92, p. 1)

- Council Directive of 3 November 1992 concerning the minimum requirements for improving the safety and health protection of workers in the mineral-extracting industries through drilling (92/91/EEC, OJEC N L348 of 28.11.92, p. 9)

- Council Directive of 3 December 1992 on the minimum requirements for improving the safety and health protection of workers in surface and under-ground mineral-extracting industries (92/91/EEC, OJEC N L348 of 28.11.92, p. 9)

- a recommendation of 22 May 1990 to the Member states concerning the adoption of a European schedule of occupational diseases (90/326/EEC, OJEC No L 160, 26.6.90, p.39) and

- a Commission directive of 29 May 1991 on establishing indicative limit values by implementing Council Directive 80/1107/EEC on the protection of workers from the risks related to exposure to chemical, physical and biological agents at work (91/322/EEC, OJEC No L 177, 5.7.91, p. 22)

The definitions used (undertakings - workers - employers) give the framework directive, and hence also the individual Directives, a very wide scope. As provided for in Article 118A, the directives:

- comprise minimum requirements

- take into account, as much as possible, small and medium-sized undertakings (the framework in its administrative provisions; the workplace and the work equipment in the amount of time allowed to adapt existing situations).

All the directives and proposals put the emphasis on information, training and consultation of workers.

More proposals are on their painful legislative way through the Institutions, as the one dealing with a Council regulation for a European Agency for safety and health at work is close to final adoption. Furthermore, a common position has been adopted for the proposal of a Council directive on health and safety on fishing vessels.

In order to complete the action programme the Commission will still have to submit more proposals which are presently in preparation in the services, such as the ones concerning specific work equipment, agriculture chemical substances and a completion of the carcinogens directive.

Legislative work which is achieved, proposed or prepared by other directorates of Social Affairs is also of interest to safety and health at work. The subjects concerned are atypical work contacts, disabled workers, young workers, ...

After this long list let me still add something very important.

All this legislative work will be worthless if it is not effectively applied in practice. Therefore the Commission has to ensure that the implementation of this legislation in the Members States has been correctly performed.

Also the means involved by the Member States to enforce this legislation must be sufficient to guarantee the application on site of the different measures. Finally, and I think this is the most important catalyst for achieving better safety and health conditions at work, a positive change in the minds of the different participants, be they employers, workers or even civil servants, is required. Safety and health shall not be an irksome accessory ranking after all other preoccupations, but it shall become a fully integrated part of all reflexions and actions in daily work.

© Commision of the European Communities

Assessment of a company for safety and minimization of losses

I NEIL, MIOSH, MIIRSM, FInstMM
Acer Consultants Ltd/Willis Corroon Risk Management Ltd, Abingdon, UK

INTRODUCTION

'Risk' in insurance terms traditionally refers to insured peril.

A seventeen year old car driver, for example, is referred to as a bad risk. This simply means that the expected claims cost is high, ie: the probability of an accident x the size of loss.

Hence premium is set to reflect this and the 'risk' has been managed by financing.

Alternatively the Risk Manager may elect to transfer the risk by means of re-insurance, thus protecting the undertaking from loss.

In simple terms the losses were either financed or transferred, with any dispute settled via the medium of the civil court of law.

Companies, could therefore offset potential losses by insuring themselves against the loss to cover such matters as:

- Professional indemnity
- Product liability
- Public or Third party liability
- Employee liability
- Business Interruption
- Contractors all risks
- Directors and Officers
- Error and Emission
- Motor vehicle
- Fire
- Legal Expenses

When assessing the company the 'Risk Assessor' would see that such cover and the level of cover were appropriate and sufficient to minimise potential foreseeable loss and in certain instances, sudden and unforeseen events.

However this is not sufficient to cover "Total Risk" or "Total Loss" and in assessing a company and its management for safety and minimisation of loss then there is a need to examine.

a) Risk arising from breach of statutory duty, in particular Health and Safety Legislation.

b) Uninsured and uninsurable losses arising from accidents in the workplace.

Risk From Breach Of Statutory Duty

This can arise from:

- Fines
- Imprisonment
- Disqualification of Directors
- Legal Costs

Commenting on the Annual Report of the Health and Safety Executive. The Director General Mr John Rimmington stated "Senior Managers will be charged individually" if we can connect top executives to blood on the he floor?"

Liability under the Health and Safety at Work etc Act 1974 increases the Risk to managers.

Liability Under Statute

Section 33 of the Act makes it an offence for an employer company or supplier, etc to fail to discharge a duty prescribed under the Act, for example by being unaware of and thus not complying with regulations made under the Act. The section goes on to provide that a person guilty of such an offence will be liable on summary conviction to a fine not exceeding £20,000. More serious offences which are dealt with in the Crown Court may involve liability for a much larger fine (there is no statutory limit) and/or up to 2 years imprisonment.

Section 37 of the Act provides that where an offence has been committed by a body corporate "with the consent or connivance of, or to have been attributable to any neglect on the part of, any director, manager, secretary or other similar officer of the body corporate or a party who was purporting to act in such a capacity, <u>he, as well as the body corporate,</u> shall be guilty of that offence and shall be liable to be proceeded against and punished accordingly."

The spirit of this legislation reflects a judgement of Lord Denning in H. L. Bolton (Engineering) Co Ltd v T. J. Graham and Sons Ltd [1957] 1 QB 159 where he observed:-

"A company may in many ways be likened to a human body. It has a brain and nerve centre which controls what it does. It also has hands which control the tools and act in accordance with directions from the centre. Some of the people in the company are mere servants and agents who are nothing but hands to do the work and cannot be said to represent the mind or will. Others are directors or mangers who represent the directing mind and will of the company, can control what it does. <u>The state of mind of these managers is the state of mind of the company and is treated by the law as such</u>".

Offences giving rise to imprisonment in the Act are fairly narrow and include:

- acquiring, possessing or using explosives in contravention of any relevant statutory provisions,

- contravening a licensing requirement, and

- disobeying a prohibition notice.

The company Directors Disqualification Act 1986 contains a provision which allows the court to make a disqualification Order against a person convicted of an indictable offence connected with the promotion, formation, management or liquidation of a company. The conduct at which the legislation was aimed may not simply be financial. The act may apply in respect of health and safety matters.

On 28th June 1992 at Lewes Crown Court (unreported) Rodney James Chapman became the first director to be disqualified in connection with this sort of crime. Mr Chapman was prosecuted by the Health and Safety Executive under Section 37 of the Health and Safety at Work, etc, Oct. 1974. In this case Mr. Chapman's company had contravened a prohibition notice which had been served on Chapman Chalk Supplies Limited in June 1991. It was alleged that the quarry in question had been worked in an unsafe manner resulting in significant danger from falls of rock. Following an appeal against the prohibition notice an industrial tribunal had ordered that the notice would only be lifted on condition that the Health and Safety Executive was satisfied with the steps which had been taken to make the quarry safe. However, employees returned to work on 12th August 1991 without the prior approval of the Executive and thus the Company contravened not only the Tribunal order and Section 33. Mr.. Chapman was fined £5,000 and the company was fined for the same amount and ordered to pay costs of £3,533. Mr. Chapman was banned from being a Company Director for two years.

In addition we have seen "Risk" to Directors extended to Criminal Negligence Manslaughter.

On 1st December 1989 Mr. Norman Holt director of a Lancashire plastics company became the first manager convicted of manslaughter arising from a breach of the Act. These charges followed the death in may 1989 of a twenty five year old worker. The man died when he was drawn into a shredding machine into which he was feeding plastic sheeting. The guard had been removed under the instruction of Mr N.Holt ,a director. The incident occurred despite instructions from H.M.Inspector of Factories to re-instate the guard to its proper condition and an undertaking from Mr Holt that it would not be removed in the future. Other charges ,including failing to register a factory,failing to securely fence a machine,breach of a prohibition notce were brought against the Directors nameley Norman and David Holt. Norman Holt was convicted of manslaughter after pleading guilty and was sentenced to 12months imprisonment suspended for two years.David Holt pleaded not guilty and the trial judge directed that the manslaughter charge be left on file.In addition the brothers were fined £5000 and £15500 respectively and the company were fined £26500 plus £10000 costs.

Also the crime of 'corporate manslaughter' is now recognised in British Law although no actual convictions have taken place to date.The Herald of Free Enterprise disaster lead to the parent company P&O being charged with corporate manslaughter.Although the case was successfully defended it,nevertheless,established the feasability of corporate manslaughter in British law and future cases may succeed if it can be shown that the directing mind of a corporate body behaved recklessly.

Penalties for breach of statutory duty can be very significant indeed and have a major impact on company profitability.

A contractor was fined £160,000 & £28,000 costs when a workman was electrocuted whilst moving an electrical transformer.

BP were fined £750,000 following an explosion at their Grangemouth refinery.

A national electricity generating company was fined £10,000 plus costs of £28,568 following the death of a contractors welder due to an explosion.

Whilst the costs may be recoverable from insurers. The fines are an uninsurable cost.

In each case listed above the companies concerned, if making 10% profit would need to generate additional sales of

£1,600,000; £7,500,000 and £100,000 respectively to maintain their profitability.

Cost of Accidents at Work

The cost of accidents can be broken down into four categories as shown below

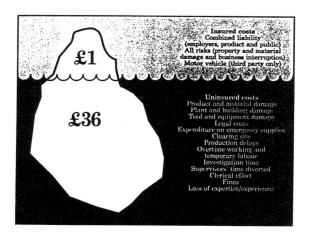

In their publication "The Costs of Accidents at Work". The Health and Safety Executive examine five case studies of the accident experience across diverse industries. Their findings indicate that for every £1 recoverable cost there was up to £36 non recoverable costs. As shown in Fig. 2 and 3 (reproduced by kind permission of the Health and Safety Executive and Her Majesties Stationary Office).

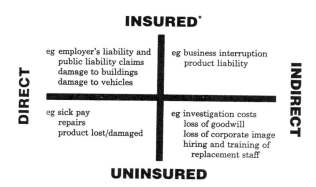

* For a particular company the amount and type of insurance cover held will vary. For example, if only third party vehicle insurance in held, all own vehicle damage will be uninsured cost.

Figure 3 Summary of losses identified *

	Total loss	Annualised loss	Representing
1 Construction site	£245.075	£700.000**	8.5% of tender price
2 Creamery	£243.834	£975.336	1.4% of operating costs
3 Transport company	£48.928	£195.712	1.8% of operating costs
4 Oil platform	£940.921	£3,763.684	14.2% of potential output
5 Hospital	£99.285	£397.140	5% of annual running costs

*   Figures quoted are actual at time of study: no adjustment has been made for inflation. Study 1 lasted 18 weeks; studies 2-5 13 weeks.

**   Represents total length of contract.

Actual figures showing the cost of accidents to British industry per annum are difficult to obtain but estimates vary from £800 million to £1.7 billion per annum.

The costs of illness and disease, arising directly from work activities is inestimable.

Organisations which manage health and safety successfully will therefore:

  Minimise their Risk

  Reduce their indirect costs

  Improve their claims history thus reduce or stabilise premiums

  Protect their resources

  Increase profitability

Assessing the Company for Safety

A successful company will have a clearly defined Health and Safety Action Plan.

Health and Safety will be managed via the line management structure with the same priority as financial planning, product planning, programming and skills training.

Central to the plan will be the company health and safety policy document which will show clearly the commitment of the Board of Directors, the command structure for implementing the policy, the training requirements for all level of employee across the company.

There should be a clear statement of the corporate health and safety objectives.

Detailed plans on how these are to be achieved.

Performance standards need to be set to measure, review and audit the policy.

The Managing Health and Safety at Work Regulations 1992 require companies to manage health and safety.

Reg 3 requires every employer to make a suitable and sufficient assessment of

a) the risks to the health and safety of his employees to which they are exposed whilst they are at work; and

b) the risks to the health and safety of persons not in his employment arising out of or in connection with the conduct by him of his undertaking for the purpose of identifying the measures he needs to take to comply with the requirements and prohibitions imposed upon him by or under the relevant statutory provisions.

This will be the criteria under which management will be assessed in future.

Those who measure up, shall prosper. Those who don't may have difficulty obtaining any form of "Risk" cover and will inevitably fail.

Quality risk and safety

F WARNER, BSc, FRS, FEng, FIMechE
Department of Chemistry and Biological Chemistry, University of Essex, Colchester, UK

SYNOPSIS There is no such thing as absolute safety. The aim of professional engineers is to reduce risk, whether this arises from design, construction or operations. The background is set down in the definitions used in BS 4778, forming part of the standards in Handbook 22 of BSI which culminate in BS 5750 on Quality Assurance and BS 7570 on Total Quality Management. The use of these gives a framework for decision based on the statistics for availability, reliability and maintenance. These are at the roots of quantitative risk assessment (QRA) through fault tree analysis, the checks by customers and independent auditors as developed in systems like hazop.

1 INTRODUCTION

1992 was the European Year of Health and Safety, marked towards its conclusion by a 5-day international conference. The complete proceedings and discussions have been published in 2 volumes by the Health and Safety Executive and the Health and Safety Directorate of the European Community. These provide a wealth of information on current practices throughout the Community with an emphasis throughout on the role of quantitative risk assessment. It is the purpose of this paper to discuss the importance of the procedures and to look at the bases on which risk assessment is carried out.

It is necessary at the beginning to be sure that terms are being used in the same way - at any rate by practising engineers. The public appreciation of risk is clouded by the inexact use of terms and unwillingness to define them. In some academic circles, there is a refusal to accept any definitions but to say that the only meaning put on risk by the public is a threat.

In engineering and scientific circles, there has been much discussion directed to definitions which are regarded as soundly based. A Royal Society meeting (1981), stated:- 'The techniques of risk assessment go back to the need for reliability in military equipment and air transport. They can now use failure rates held in data banks containing rapidly increasing amounts of information. The use of fault-trees in risk analysis gives a logical basis for reducing risk during the conceptual state of projects. A number of papers deal with the problems of assessment shown up in conventional, unconventional, and nuclear power plants. The problems have existed much longer in the risks of failure in bridges, dams, buildings, chemical and petroleum installations, and transport. Risks associated with drugs and medical procedures are complicated by the benefits weighed against them; the risks also show up only over long periods as a result of epidemiological studies and finally in mortality tables. The papers discuss not only the risk that is final, that of death, but also of injury up to fates worse than death, with special references to risk at work' (1).

This meeting led to formation of a Study Group which reported in 1983 (2) and in para 1.6 on Terms used stated: For the purposes of this report the Study Group views RISK *as the probability that a particular adverse event occurs during a stated period of time, or results from a particular challenge.* As a probability in the sense of statistical theory risk obeys all the formal laws of combining probabilities. Explicitly or implicitly, it must *always* relate to the 'risk of ... (a specific event or set of events)' and where appropriate must refer to an exposure to hazard specified in terms of its amount or intensity, time of starting or duration. The word 'risky' is undefined, and is *not* to be used as a synonym for 'dangerous'. All risks are conditional, although often the conditions are implied by context rather than explicitly stated. The risk of death while hang-gliding during a seven-day period is small for a randomly selected inhabitant of the UK, but its value will alter substantially according to age, season, weather and membership of the hang-gliding club.

An *ADVERSE EVENT is an occurrence that produces harm* (shortened to 'event' where unambiguous).

With risk defined as above, HAZARD is seen as *the situation that in particular circumstances could lead to harm*, where HARM is *the loss to a human being (or to a human population) consequent on damage* and DAMAGE is *the loss of inherent quality suffered by an entity (physical or biological).* BENEFIT is *the gain to a human population.* Expected benefit incorporates an estimate of the probability of achieving the gain.

Consider the existence of Nelson's Column as the hazard, It may be damaged by wind and lightning and, as a consequence, pieces may fall off and cause harm to people in Trafalgar Square. Risk would measure the probability of specified damage or harm in a given period.

DETRIMENT is a *numerical measure of the expected harm or loss associated with an adverse event,* usually on a scale chosen to facilitate meaningful addition over different events. It is generally the integrated product of risk and harm and is often expressed in terms such as costs in £s, loss in expected years of life or loss in productivity, ad is needed for numerical exercises such as cost-benefit or risk-benefit analysis. Although detriment may represent the only numerical way of comparing different events associated with the same hazard, or the combined effects of events from different hazards, the fact that any such comparison is an arbitrarily weighted total of incommensurables must never be forgotten. Total detriment (per individual or per population) may be an aid to decision, especially when reasonable alternative systems of weighting lead to the same conclusion, but should not be regarded as a substitute for reasoned judgment.

The general term used to describe the study of decisions subject to uncertain consequences is RISK-ASSESSMENT. It is conveniently sub-divided into *RISK-ESTIMATION* and *RISK-EVALUATION*. The former includes:

(a) the identification of the outcomes;
(b) the estimation of the magnitude of the associated consequences of these outcomes; and
(c) the estimation of the probabilities of these outcomes.

RISK-EVALUATION is the complex process of determining the significance or value of the iden-

tified hazards and estimated risks to those concerned with or affected by the decision. It therefore includes the study of risk perception and the trade-off between perceived risks and perceived benefits. *RISK-MANAGEMENT is the making of decisions concerning risks and their subsequent implementation*, and flows from risk-estimation and risk-evaluation.

For engineers, the definitions used are found in the corpus of standards relating to quality which comprise Handbook 22 of the British Standards Institution. They spell out in more detail the generalizations of the Royal Society report and appear in Section 2 of BS 4778 1991 as follows: The concept of risk is embedded in quality assurance. The Royal Society Report (1992) *Risk - analysis, perception and management* (3) defines risk in terms of the probability that a particular adverse event occurs during a stated period of time, or results from a particular challenge. This has a parallel in the definition given in (c) 2 of *BS 4778* (4) which appears in Handbook 22 as part of the corpus of standards to be used along with *BS 5750* (5). The International Electrotechnical Commission adopts all the definitions of direct relevance to this Lecture and they are translated in BS 4778 and, in this form, appear in IEC50 (1991) of the in German, French, Russian and Spanish as well as English. Apart from the definitions, BS 4778 and the other standards in Handbook 22 lay the basis for the concepts and statistical methods which are the tools of quality assurance (QA). They reach their general statement in BS 5750 which has found international acceptance as ISO 9000 and in Europe as EN 29000.

Since the publication of Handbook 22, two further standards have appeared which are relevant to quality and also to risk where the environment is concerned. BS 7750 is the standard for environmental auditing and BS 7850 gives the most up-to-date guidance on Total Quality Management (TQM).

For engineering purposes, the first step in TQM is to be sure that data are recorded. The information needed is summarized under 3 heads - Availability, Reliability and Maintenance (ARM). It is to be found as the first item in Table 1 of BS 7850:1992 Part 1 on selecting an available tool or technique for the purposes of TQM.

This includes Pareto diagrams for establishing a rank order for dealing with quality problems. It is a special example of what will be discussed later - risk/benefit analysis for deciding how to use resources which are not unlimited.

Total quality management, and quality assurance as part of it, is directed to getting the best result in terms of goods or services from the human and material resources available. Table 1 shows the steps which are needed for selecting appropriate techniques. They have their origin in manufacturing processes and the way statistics are used to achieve the quality required. It is followed by the statement " ... there is a common need to ensure that management responsibilities for the processes, especially those that flow across occupational boundaries, are clearly defined including those for health and safety ... ". It is in this area where the statistics show that risk in terms of probability of an adverse event is at a higher level than is attainable in the case of product failure.

There has been criticism of BS 5750 as bureaucratic and unsuitable for small firms. There is misunderstanding of the meaning of certification to this standard. It is not a product standard. These are laid down in their separate and individual specifications. It is a standard for performance to ensure that the required quality is met and is basically a system of management which provides information and checks at all levels of operation in an organization so that the performance required corresponds with the level of competence and understanding.

Although primarily directed to enhancing the operations of an organization, QA has the components which enhance safe operation through the understanding of the processes which are being carried out. The introduction underlines that there is no such thing as absolute safety but there are figures for risk which can be reduced by attention to the details of design, construction and operation. The simplest criterion of risk is that of death over a specified period. Some risks are given in Table 2.

The immediate reactions to this Table are to note that the average risk of death is greater than $10^{-2}y^{-1}$ (taking the population of the UK around 50 millions for ease in estimating). The risk of death from a road accident is slightly less than $10^{-4}y^{-1}$. There are a number of details in this risk which show up factors which influence decisions relating to risk reductions. The present figure for all ages shows a reduction from a figure of $1.22 \: 10^{-4}y^{-1}$ prior to the compulsory wearing of seat-belts. In easily comprehensible figures (again using the hypothetical population of 50 millions) this represents a reduction of 1200 deaths a year from the earlier total of 6000. It illustrates a problem in the perception of risk. Legislation to make wearing of seat-belts compulsory was passionately opposed in parliamentary debates on the grounds of interference with individual liberty.

The overall figure of risk from road accidents is made up of the different figures according to age group. The age group 10-14 has only about half the overall risk where 15-19 has 2.5 times the risk, coming down to twice for 20-24 before coming down to just below the overall risk for the 25-29 age group.

The risks given in Table 2 are based upon the statistics available in data bases of general application (Table 3).

The examination of risks arising from sporting activities is difficult because the time base is not easy to establish. It is important to attempt a comparison, since these risks are voluntarily assumed and fall into the category of acceptability, the improvement in quality of life from the excitement of undertaking the sport being valued more than the potential loss of life expectancy.

Table 4 gives some figures.

An attempt to compare risks at work on the basis of the time spent in a working life was pioneered over 20 years ago in the chemical industry. It used a fatal accident rate (FAR) which assumed a total working life of 10^8 hours and proved a powerful tool for purposes of comparison. It was used alongside hazard and operability studies (HAZOP) which carried out on paper the construction of fault trees, using ARM data. These constituted a proactive study for the recognition and elimination of high risks in a procedure which combined the skills of designers, builders and operators with an audit operation included.

The chemical industry was brought under a spot-light by the disaster at Flixborough in 1974 which occurred as the Health and Safety at Work Act was passed. It led to the Committee on Major Hazards which carried out many studies on setting up Quantitative Risk Assessment. The reduction in risk at work is shown in Table 5.

From this the chemical industry compares well with others. It is ironic that it should be singled out for major hazards when the risks in fact are low and progressively reduced.

Much of this reduction is due to the Health and Safety Executive through its policy of education as much as regulation. In communicating with the public at large and the workforce in particular, HSE has carried out a number of studies related to perception and attitudes. Its recent publication (The Tolerability of Risk from Nuclear Power Stations, (14)) tried to illustrate the areas of concern by use of f(N) curves. Figure 1 shows the suggested boundaries for tolerable risk informed by the use of the ALARP (as low as reasonably practicable) principle. The Figure plots frequency of N or more fatalities against the number N on a log-log scale. The Royal Society Risk Report (3) has put together a number of f(N) curves (Figure 2). The solid lines are plotted from historical data, the dotted lines from risk estimation. They are for man-made, not natural, hazards. The chief natural hazard is flooding, in which deaths of 100 000 upward are almost an annual occurrence, and up to a million possible. These curves have special interest at the present time following the Mississippi floods and monsoon flooding in N. India.

An area of particular interest to engineers is transport and the risks are expressed in one useful form in terms of 10^9 km travelled. Table 6 shows the advantage of public over private transport in respect of risk. If rail transport in the UK is considered, the risk in conventional terms is between $10^{-7}y^{-1}$ and $10^{-8}y^{-1}$ since roughly a major accident in which 100 people are killed occurs roughly every 10 years. It is limited by the number of people normally travelling in a train — low in this country but high in others. The limitations because of container size is relevant to other forms of transport. It can even be considered as a more general problem of geometry. In estimating the potential risks of fire, explosions or release of toxic gases, the concentration of material which is hazardous is largely determined by an inverse square law.

2 RISK BENEFIT ANALYSIS

The decisions made by regulatory agencies should take account of the risk which they seek to reduce and a value to be placed on avoidance of premature death or serious injury. Very few studies have been made of the latter but a figure was taken by the Department of Transport for avoiding one death of £ 0.5 million (1988 values) and quoted in a Treasury Green paper (1991) (15). The Royal Society Report puts a value of £ 200 - £ 300 for each change in the risk of death of 1/10 000 which expressed conventionally would mean £ 2-3 million for one death avoided.

The measures taken by regulatory authorities in the USA were discussed at the international workshop on Risk Assessment (1992) by Belzer from the President's Executive Office (16).

His Table (IV) shows a column for cost per Premature Death Averted as a result of risk management decisions in $ million as 0.1 for car seat belts to 5 700 000 for wood-preserving chemicals (Table 7).

Information of this detail is not available in the U.K. but will be essential if public agencies are to meet the requirements to publish codes of conduct by January 1994 as part of DTI action to promote a de-regulation Bill. The Health and Safety Executive will be affected.

This paper has attempted to deal with some aspects of quality risk and safety and hopefully complements other contributions to this meeting.

REFERENCES

(1) WARNER, F.E. AND SLATER, D.H. The Assessment and Perception of Risk, Proc. R. Soc. Lond., 1981, A376, 1-206.

(2) ROYAL SOCIETY. Risk Assessment, 1983 (The Royal Society).

(3) ROYAL SOCIETY. Risk - Analysis, Perception and Management, 1992 (The Royal Society).

(4) BRITISH STANDARDS INSTITUTION. Quality Vocabulary - BS4778, 1991 (BSI, London).

(5) BRITISH STANDARDS INSTITUTION. Quality Systems - BS5750, 1991 (BSI, London).

(6) BRITISH STANDARDS INSTITUTION. Total Quality Management - BS7850, 1992 (BSI, London).

(7) HEALTH AND SAFETY EXECUTIVE. Mines and Quarries, 1978, Mines, 1979, Quarries, 1979 (London: HMSO).

(8) HEALTH AND SAFETY EXECUTIVE. Health and Safety Commission Report 1978-79, 1980 (London: HMSO).

(9) HEALTH AND SAFETY EXECUTIVE. Health and Safety Commission Report 1979-80, 1980 (London: HMSO).

(10) BRITISH HANG GLIDER ASSOCIATION. Personal Communication, 1981.

(11) BRITISH SUB-AQUA CLUB. Personal Communication, 1981.

(12) CLARK, K.S. Calculated Risks of Sports Fatalities. Journal of the American Medical Association, 1966, 197, 894-896.

(13) METROPOLITAN LIFE ASSURANCE COMPANY. Statistical Bulletin, 1979, 60 (3), 2.

(14) SOWBY, F.D. Radiation and Health Risks. Health Physics, 1965, 11, 879-887.

(14) HEALTH AND SAFETY EXECUTIVE. The Tolerability of Risks from Nuclear Power Stations, 1992 (London: HMSO).

(15) H.M TREASURY. Economic Appraisal in Central Government, 1991 (London: HMSO)

(16) BELZER, R.B., The Use of Risk Assessment and Benefit-Cost Analysis in US Risk-Management Decsion Making, Risk Assessment, 1992 (Health and Safety Executive).

Table 1: Selecting an appropriate tool or technique

Tool or Technique	When to Select
Data collection form	Gather a variety of data in a systematic fashion for a clear and objective picture of the facts
Tools for non-numerical data	
Affinity diagram	Organize into groupings a large number of ideas, opinions, issues, or other concerns
Benchmarking	Measure your process against those of recognized leaders
Brainstorming	Generate, clarify, and evaluate a sizeable list of ideas, problems, or issues
Cause and effect diagram	Systematically analyze cause and effect relationships and identify potential root causes of a problem
Flow chart	Describe an existing process, develop modifications, or design an entirely new process
Tree Diagram	Break down a subject into its basic elements
Tools for numerical data	
Control chart	Monitor the performance of a process with frequent outputs to determine if its performance reveals normal variations or out-of-control conditions
Histogram	Display the dispersion or spread of data
Pareto diagram	Identify major factors and distinguish the most important causes of quality losses from the less significant ones
Scatter diagrams	Discover, confirm or display relationships between two sets of data

Note: These tools are described in BS 7850: Part 2. (6).

Table 2: Some risks of death expressed as annual experience per million of the population of the U.K. for 1989

Average over entire population	11490
Men aged 55-64	15280
Women aged 55-64	9060
Men aged 35-44	1730
Women aged 35-44	1145
Boys aged 5-14	225
Girls aged 5-14	160
Death by accident (all)	240
Death by road accidents (averaged over population)	98
Death by accident in the home	86
Homicide, England and Wales 1990*	12
Homicide from terrorism, England and Wales 1982-90*	0.2

Source: Health and Safety Executive (7-9) and Consumer Safety Unit
*Home Office

Table 3: Sources of data relating to physiological status and disease prevalence and incidence

National and area vital statistics (births, deaths, and census data)

Hospital records (discharge diagnoses and operations)

General practitioner records

Sickness benefit certificates

School medical examinations

Health visitors' records

Morbidity registers and notifications (congenital anomalies, cancer, blindness, infectious diseases, adverse reactions to drugs)

Industrial medical records

General Household Survey

Ad hoc surveys

Table 4: Accidental death rates attributed to sporting activities

	Deaths per 10^6 participant-hours*
School and college football	0.3
Amateur boxing, U.K., 1946-62	0.5
Skiing, U.S., 1967-68	0.7
France, 1974-76	1.3
Canoeing, U.K., 1960-62	10
Mountaineering, U.S., 1951-60	27
Motorcycle racing, U.K., 1958-62	35
Rock climbing, U.K., 1961	40

	Deaths per 10^6 participant-years**
Cave exploration, U.S., 1970-78	45
Glider flying, U.S., 1970-78	400
Scuba diving, U.K., 1970-80	220
U.S., 1970-78	420
Hang gliding, U.S., 1978	400 to 1300
U.K., 1977-79	1500
Power boat racing, U.S., 1970-78	800
Sport parachuting, U.S., 1978	1900
Association football, England and Wales 1986-90	1.2
Climbing, England and Wales, 1986-90	130
Motor sports, England and Wales, 1986-90	27

* Based on approximate estimates of participants' hours per year spent in the activity

** Based on numbers of participants and deaths per calendar year, without allowance for hours actually spent in the activity

Sources include: British Hang Glider Association (1981) (10), British Sub-aqua Club (1981) (11), K. S. Clarke (1966) (12), Metropolitan Life Assurance Co. (1979) (13) and F. D. Sowby (1965) (14).

Table 5: Average annual accidental death rates at work in the U.K. per million at risk (1974-78 and 1987-90 except as stated)

	1974-78	1987-90
Manufacture of clothing and footwear	5	0.9
Manufacture of vehicles	15	12
Manufacture of timber, furniture, etc.	40	22
Manufacture of bricks, pottery, glass, cement etc.	65	60
Chemical and allied industries	85	24
Shipbuilding and marine engineering	105	21
Agriculture (employees)	110	74
Construction industries	150	100
Railway staff	180	96
Coal miners	210	145
Offshore oil and gas	1650*	1250
Deep sea fishing (accidents at sea)	2800**	840

*1967-76
**1959-68

Source: Health and Safety Executive (7-9)

Table 6: Deaths per 10^9 km travelled, UK

	1967-71	1972-76	1986-90
Railway passengers	0.65	0.45	1.1
Passengers on scheduled air services on UK airlines	2.3	1.4	0.23
Bus or coach drivers and passengers	1.2	1.2	0.45
Car or taxi drivers and passengers	9.0	7.5	4.4
Two-wheeled motor vehicle driver	163.0	165.0	104.0
Two-wheeled motor vehicle passenger	375.0	359.0	104.0
Pedal cyclists	88.0	85.0	50.0
Pedestrians*	110.0	105.0	70.0

* Based on a National Travel Survey (1985/86) figure of 8.7km per person per week.
Source: Department of Transport

Table 7

TABLE IV
Baseline Risks and Cost-Effectiveness of Selected Federal Risk-Management Regulations

Regulation (a)	Cost a Factor in Decision Making?	Year Issued	Health or Safety?	Agency	Baseline Mortality Risk per Million Exposed	Cost Per Premature Death Averted ($M 1990)
Unvented Space Heater Ban		1980	S	CPSC	1,890	0.1
Aircraft Cabin Fire Protection Standard	Yes	1985	S	FAA	5	0.1
Auto Passive Restraint/Seat Belt Standards	Yes	1984	S	NHTSA	6,370	0.1
Steering Column Protection Standard (b)	Yes	1967	S	NHTSA	385	0.1
Underground Construction Standards (c)		1989	S	OSHA-S	38,700	0.1
Trihalomethane Drinking Water Standards		1979	H	EPA	420	0.2
Aircraft Seat Cushion Flammability Standard	Yes	1984	S	FAA	11	0.4
Alcohol and Drug Control Standards (c)		1985	H	FRA	81	0.4
Auto Fuel-System Integrity Standard	Yes	1975	S	NHTSA	343	0.4
Standards for Servicing Auto Wheel Rims (c)		1984	S	OSHA-S	630	0.4
Aircraft Floor Emergency Lighting Standard	Yes	1984	S	FAA	2	0.6
Concrete & Masonry Construction Standards (c)		1988	S	OSHA-S	630	0.6
Crane Suspended Personnel Platform Standard (c)		1988	S	OSHA-S	81,000	0.7
Passive Restraints for Trucks & Buses: Proposed	Yes	1989	S	NHTSA	6,370	0.7
Dynamic Side-Impact Standards for Autos	Yes	1990	S	NHTSA	NA	0.8
Children's Sleepwear Flammability Ban (d)		1973	S	CPSC	29	0.8
Auto Side Door Support Standards (e)	Yes	1970	S	NHTSA	2,520	0.8
Low-Altitude Windshear Equip. & Training Standards	Yes	1988	S	FAA	NA	1.3
Electrical Equipment Standards -- Metal Mines		1970	S	MSHA	NA	1.4
Trenching and Excavation Standards (c)		1989	S	OSHA-S	14,310	1.5
Traffic Alert & Collision Avoidance Systems	Yes	1988	S	FAA	NA	1.5
Hazard Communication Standard (c)		1983	S	OSHA-S	1,800	1.6
Lockout/Tagout (c)		1989	S	OSHA-S	4	2.1
Side-Impact Stds for Trucks, Buses & MPVs: Proposed	Yes	1989	S	NHTSA	NA	2.2
Grain Dust Explosion Prevention Standards (c)		1987	S	OSHA-S	9,450	2.8
Rear Lap/Shoulder Belts for Autos	Yes	1989	S	NHTSA	NA	3.2
Benzene NESHAP [Original: Fugitive Emissions]		1984	H	EPA	1,470	3.4
Standards for Radionuclides in Uranium Mines (c)		1984	H	EPA	6,300	3.4
Ethylene Dibromide Drinking Water Standard		1991	H	EPA	NA	5.7
Benzene NESHAP (Revised: Coke By-Products) (c)		1988	H	EPA	NA	6.1
Asbestos Occupational Exposure Limit (c)		1972	H	OSHA-H	3,015	8.3
Benzene Occupational Exposure Limit (c)		1987	H	OSHA-H	39,600	8.9
Electrical Equipment Standards -- Coal Mines (c)		1970	S	MSHA	NA	9.2
Arsenic Emission Standards for Glass Plants		1986	H	EPA	2,660	13.5
Ethylene Oxide Occupational Exposure Limit (c)		1984	H	OSHA-H	1,980	20.5
Arsenic/Copper NESHAP		1986	H	EPA	63,000	23.0
Haz Waste Listing for Petroleum Refining Sludge		1990	H	EPA	210	27.6
Cover/Move Uranium Mill Tailings, Inactive Sites		1983	H	EPA	30,100	31.7
Benzene NESHAP [Revised: Transfer Operations]		1990	H	EPA	NA	32.9
Cover/Move Uranium Mill Tailings, Active Sites		1983	H	EPA	30,100	45.0
Acrylonitrile Occupational Exposure Limit (c)		1978	H	OSHA-H	42,300	51.5
Coke Ovens Occupational Exposure Limit (c)		1976	H	OSHA-H	7,200	63.5
Asbestos Occupational Exposure Limit (c)		1986	H	OSHA-H	3,015	74.0
Arsenic Occupational Exposure Limit (c)		1978	H	OSHA-H	14,800	106.9
Asbestos Ban	Yes	1989	H	EPA	NA	110.7
Diethylstilbestrol [DES] Cattlefeed Ban		1979	H	FDA	22	124.8
Benzene NESHAP [Revised: Waste Operations]		1990	H	EPA	NA	168.2
1,2-Dichloropropane Drinking Water Standard		1991	H	EPA	NA	653.0
Hazardous Waste Land Disposal Ban ["1st 3rd"]		1988	H	EPA	2	4,190.4
Municipal Solid Waste Landfill Standards, Revised		1991	H	EPA	<1	41,250.0
Formaldehyde Occupational Exposure Limit (c)		1987	H	OSHA-H	31	86,201.8
Atrazine/Alachlor Drinking Water Standard		1991	H	EPA	NA	92,069.7
Hazardous Waste Listing for Wood Preserving Chemicals		1990	H	EPA	<1	5,700,000.0

Agency Abbreviations:
CPSC = Consumer Product Safety Commission
EPA = Environmental Protection Agency
FAA = Federal Aviation Administration, DOT
FDA = Food and Drug Administration, HHS
OSHA-H = Occupational Safety and Health Administration, Health Standards, DOL
OSHA-S = Occupational Safety and Health Administration, Safety Standards, DOL
MSHA = Mine Safety and Health Administration, DOL
NHTSA = National Highway Traffic Safety Administration, DOT
FRA = Federal Railroad Administration, DOT

Notes:
(a) 70-year lifetime exposure assumed unless otherwise specified.
(b) 50-year lifetime exposure assumed.
(c) 45-year lifetime exposure assumed.
(d) 12-year exposure assumed.

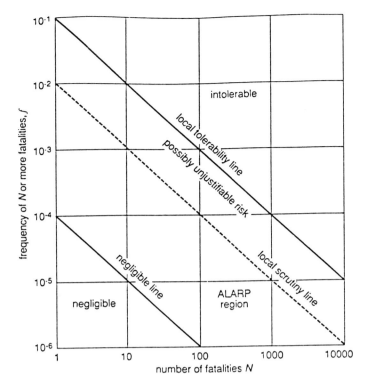

Figure 1

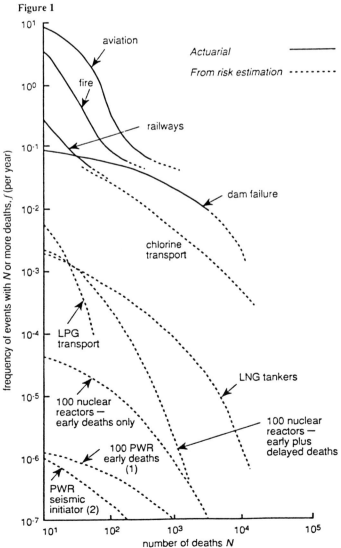

Figure 2. Examples of f vs N lines for various man-made hazards

The role of human factors and safety culture in safety management

R T BOOTH, PhD, DIC, CEng, FIMechE, FIOSH, RSP
Health and Safety Unit, Department of Mechanical and Electrical Engineering, Aston University, Birmingham, UK

T R LEE, MA, PhD, CPsychol, FBPsS
Environmental Psychology and Policy Unit, School of Psychology, University of St Andrews, Fife, UK

SYNOPSIS The paper describes the evolution of safety management, and the part played by human factors in accident causation. It identifies the key elements of effective safety management, and suggests that a crucial determinant of good safety performance is the safety culture of the organisation. A plan is outlined for reviewing and improving safety culture with the support of a detailed safety culture promptlist.

1 INTRODUCTION

The UK is experiencing in the last decade of the 20th Century the most fundamental changes in health and safety regulation than perhaps at any time since the introduction of the first safety legislation in the mid 19th Century. The new regulations, and in particular the Management of Health and Safety at Work Regulations 1992, build on the skeleton framework for safety management contained in the Health and Safety at Work Act 1974.

In parallel with these legal developments, substantial progress has been made in identifying and classifying the crucial ingredients of effective safety management, and the approach companies should adopt to improve their safety performance (1). The new safety management regulations underpin contemporary good practice. It is now clear that safety management and corporate management generally should share a common approach. The *safety culture*, a sub-set of the overall organisational culture, is now believed to be a key predictor of safety performance (2). The safety culture of an organisation is essentially a description of the attitudes of personnel about the company they work for, their perceptions of the magnitude of the risks to which they are exposed, and their beliefs in the necessity, practicality, and effectiveness of controls. Indeed, aspects of the overall culture (for example, the effectiveness of communications generally in an organisation) may have more influence on accident rates than many elements of mainstream safety programmes (3).

A series of major accidents in transportation and in off shore oil exploration in the late 1980's demonstrated the disastrous consequences of weaknesses in safety organisation and culture. A significant finding of the inquiry reports (4 to 7) was that senior managers shared a mistaken belief that they were working in organisations with total commitment to safety and with effective safety systems. The managers may not have known how to seek out, or to recognise, the symptoms of an unsafe organisation. It may also be true that they would not have known what practical steps to take in order to turn an unsafe into a safe organisation. The inquiry reports emphasised the priority that enterprises must attach to prediction and measurement of safety performance. The accidents reveal that the behaviour of personnel whose errors may lead directly to harm, may be strongly influenced by the decision (latent) failures by more senior staff. Intentional violations of procedures by staff at all levels may pose at least as great a threat to the safety of the operation as unintended errors.

The paper is based on *Organising for safety* (2) prepared by the Human Factors Study Group of the Advisory Committee on the Safety of Nuclear Installations. We gratefully acknowledge the substantial contribution of the Chairman, the late Dr Donald Broadbent CBE FRS, and our colleagues on the Study Group, to the views presented here.

2 OBJECTIVES

The objectives of the paper are to:

(a) describe the evolution of proactive safety management and the concept of safety culture;
(b) review the part played by human factors in accident causation and prevention;
(c) discuss the crucial importance of a positive safety culture in the achievement of an effective safety management system;
(d) outline the elements of a company plan to review and promote a positive safety culture, and the use of a safety culture promptlist to assist in the development of the plan.

3 THE EVOLUTION OF SAFETY MANAGEMENT

Accident prevention requires the creation, and maintenance, of a safe working environment, and the promotion of safe behaviour - the avoidance of error - by people working with hazards. But safety management effort has traditionally been directed at the prevention of *repetitions* of accidents that have already occurred, largely on the basis of information derived from detailed accident investigations. A simplified view of the traditional approach is shown in Fig 1, derived from (8).

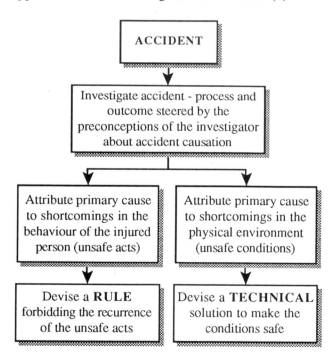

Fig 1 The traditional approach to safety management

Accidents meriting investigation usually involve a casualty, and it is not surprising that the behaviour of those who are injured may dominate the minds of the investigators. But the main reason why safety management has concentrated on reactive prevention is simply that it is a good deal easier than proactive prevention. Assessing risks and devising preventive plans without the help of accident data is difficult: it involves weighing the probabilities of a wide range of unwanted outcomes, and preparing an integrated control plan to cope with all the hazards detected. In contrast, preventing the recurrence of a particular accident is a self-contained problem with an apparently self-evident solution.

Key features of the traditional approach, based on early models of accident causation, were:

- the search for a *primary* accident cause, and,
- the debate whether the primary cause was an *unsafe act* or an *unsafe condition*.

Most practical accident prevention involved the preparation of a safety rule designed to prevent a recurrence of the unsafe act, or a physical safeguard to remedy the unsafe condition, most proximate to the accident. Much of the corpus of traditional UK health and safety legislation contains rules and physical standards derived as have been described. A satisfactory feature of the introduction of the new regulations referred to above is the phased repeal of much archaic safety law.

The causation debate, clouded by sterile political overtones and a desire to apportion blame, has often missed three crucial, and inter-related, issues:

- the concept of a single primary accident cause is a bizarre simplification of a complex multi-causal process. Moreover the term unsafe act embraces a wide range of human errors, including both unintentional errors and intentionally risky behaviour (violations);
- the distinction between the contribution of unsafe conditions and unsafe acts in *causation* has masked the more important distinction between the relative contribution of conditions and behaviour in *prevention*, and the need for prevention plans to promote both safe conditions *and* safe behaviour (9);
- the argument has focused almost exclusively on the errors made by the people who have had the accidents, not the managers and engineers whose errors (remote in time and place from the location of an accident) may have created a climate and a physical environment where errors by people at risk are made more likely or more serious. Unsafe acts create conditions where further unsafe acts may lead to accidents. The remote errors by managers have been described by Reason (10) as *latent* or *decision* failures, and the errors by people directly at risk as *active* failures.

Whatever the merits - and achievements - of the safety management approach just described, it was clearly inadequate to cope with major hazards in rapidly developing technologies. Here preventive measures may be rendered obsolete by each technical advance, and the occurrence of the first accident may itself be intolerable. But the weakness of the method is not confined to high-risk and rapidly-advancing technologies. The essential point is that rules and safeguards devised in the aftermath of disagreeable accidents may be over-zealous (as perceived some years after the accident), and conflict with the needs of both employers and employees to get the job done. Both parties may tacitly conspire to evade the safety rules or to defeat the physical safeguards. Moreover the measures taken to prevent one specific accident may conflict with the measures adopted to prevent a different accident, and with production-oriented rules. Rule books (and indeed legislation) drawn up in this way are likely to become in time both incomprehensible and contradictory. At least two company rules may exist for any situation: the rule to get the job done in time, and a more demanding safety rule that may be invoked when things go wrong.

The Robens Committee (11), whose report created the framework and philosophy of the Health and Safety at Work Act 1974, described in detail the shortcomings of safety management as it had evolved in UK industry. Robens advocated *self regulation:* the doctrine that competent and committed employers, in consultation with their workforce, would identify hazards, assess risks, and implement preventive measures, within a framework of law and standards developed nationally with the participation of all parties. Traditional prescriptive law promoted, at best, grudging and unthinking compliance. The idea of the new law was to promote proactive safety management, and enthusiastic compliance by everyone involved. We believe that the Management of Health and Safety at Work Regulations 1993, taken together with the phased repeal of traditional safety law, offers a new chance to industry to develop self regulation as envisaged by Robens.

4 ACCIDENT CAUSATION AND PRO-ACTIVE SAFETY MANAGEMENT

Accident prevention programmes must address the following distinctive elements of the accident causation process:

(a) Multi-causality

Very few accidents, particularly in large organisations and complex technologies, are associated with a single cause. Rather accidents happen as a result of a chance concatenation of many distinct causative factors, each one necessary but not sufficient to cause a final breakdown (10). It follows that the coverage of prevention plans should seek to permeate all aspects of the organisation's activities.

(b) Active and Latent failures

Active failures are errors which have an immediate adverse effect. In contrast, latent failures lie dormant in an organisation for some time only becoming evident when they combine with local triggers. The triggers are the active failures: unsafe acts, and unsafe conditions. The recognition of the importance of latent failures is useful because it emphasises the role of senior managers in causation, and draws attention to the scope for detecting latent failures in the system well before they are revealed by active failures.

(c) Skill, rule and knowledge based errors, and violations

The standard framework for classifying error is the skill- rule- and knowledge-based model proposed by Rasmussen (12) and described in (13).

Skill-based errors involve 'slips' or 'lapses' in highly practised and routine tasks. At a rather more complex level a person has to look at a situation and classify it into a familiar category as the basis for action; if it is mis-classified, this may be called a rule-based error, or mistake. Knowledge-based errors describe the most complex cases where people fail to create an adequate new rule to cope with a situation. Violations, sometimes referred to as 'risk taking', comprise a further category of error. Here, a person deliberately carries out an action that is contrary to a rule, such as an approved operating procedure.

The success of training programmes depends on an adequate diagnosis of the nature of the errors likely to be made. For example, task analysis and training which fails to consider violations may prove wholly ineffective.

(d) Hazard identification, risk assessment, and preventive action

The need to identify hazards, assess risks, and select, implement and monitor preventive actions is an essential foundation of safety management - the avoidance of latent failures. It is also the foundation for safe personal behaviour in the face of danger - the avoidance of active failures.

To create and maintain a safe working environment, and to work safely in a dangerous environment people must have the knowledge and skills and must know the rules, and be motivated, to (14):

- identify hazards;
- assess accurately the priority and importance of the hazards (risk assessment);
- recognise and accept personal responsibility for dealing with the hazards in an appropriate way;
- have appropriate knowledge about what should be done (including specified rules);
- have the skills to carry out the appropriate necessary sequence of preventive actions, including monitoring the adequacy of the actions, and taking further corrective action.

The organisation should be aware of circumstances where managers, supervisors, and other personnel may:

- underestimate the magnitude of risks;
- overestimate their ability to assess and control risks;
- have an impaired ability to cope with risks.

4.1 The aims of safety management

The primary aim of safety management is to intervene in the accident causation process and to break the causation chain. This involves preventing or detecting latent and active failures in the continuing process of hazard identification, risk assessment, control, and monitoring. However the aim of safety management is not limited simply to hazard identification, control and monitoring. Employers must plan for safety. Decisions have to be made, for example, about priorities for resource allocation, about training needs, about the appropriate risk assessment methodologies to be adopted, about the need for human reliability assessment, and about the choice of tolerable risk criteria. Safety criteria should underpin every decision made by the enterprise. Safety must be considered as an integral part of day-to-day decision-making. Moreover an employer must establish organisation and communications systems which facilitate the process of integrating safety within the management process, and which ensure that everyone in the organisation is at least fully informed of safety issues, and ideally has had an opportunity to contribute to the debate.

4.2 The key functions of safety management

From the foregoing, the four key functions of the management of safety may be summarised as follows:

(a) policy and planning

determining safety goals, quantified objectives, and priorities, and a programme of work designed to achieve the objectives, which is then subject to measurement and review.

(b) organisation and communication

establishing clear lines of responsibility and two-way communications at all levels;

(c) hazard management

ensuring that hazards are identified, risks assessed, and control measures determined, implemented, and subject to measurement and review;

(d) monitoring and review

establishing whether steps (a), (b) and (c) above:
- are in place;
- are in use, and
- work in practice.

The four key elements of safety management are underpinned by the requirements of the Management of Health and Safety at Work Regulations 1992.

4.3 Safety management and safety culture

The procedures and systems described above are necessary elements of an effective safety programme. But they are not the whole story. The danger exists that an organisation's safety policies, plans and monitoring arrangements which appear, on paper, to be well-considered and comprehensive may create an aura of respectability which disguise sullen scepticism or false perceptions among opinion formers at management and shop floor levels. The critical point is not so much the adequacy of the safety plans as the perceptions and beliefs that people hold about them. The next section focuses on the issues that determine whether the safety procedures just described are implemented with the full and enthusiastic support of the whole workforce, or whether the procedures are, at best, put into practice grudgingly and without thought or, at worst, are honoured in the breach.

5 THE CONCEPT OF SAFETY CULTURE

5.1 Origins and definition of safety culture

The concept now known as safety culture was foreshadowed in a seminal paper by Zohar (15) who described the elements of the 'safety climate' of organisations. The term safety culture was introduced to the nuclear safety debate by the International Nuclear Safety Advisory Group of the International Atomic Energy Agency (IAEA) (16) in its analysis of Chernobyl. The Agency has subsequently published an authoritative report which elaborates the concept in detail (17). The Agency (ibid) has defined safety culture as:

'... that assembly of characteristics and attitudes in organisations and individuals which establishes that, as an overriding priority, nuclear plant safety issues receive the attention warranted by their significance.'

They go on to say:

'A second proposition then follows, namely that *such matters are generally intangible*; that nevertheless *such qualities lead to tangible manifestations*; and that a principal requirement is the development of means to *use the tangible manifestation to test what is underlying*.' [our italics]

The CBI (18) describes the culture of an organisation as 'the mix of shared values, attitudes and patterns of behaviour that give the organisation its particular character. Put simply it is "the way we do things round here"'. They suggest that the 'safety culture of an organisation could be described as the ideas and beliefs that all members of the organisation share about risk, accidents and ill health'.

A possible shortcoming of the IAEA definition is that they use the term to describe only an ideal safety culture. The CBI's reference to shared ideas and beliefs does not make explicit the need for shared action. Neither definition quite captures the necessary elements of competency and proficiency. We have suggested (2) the following as a working definition:

> 'The safety culture of an organisation is the product of individual and group values, attitudes, competencies, and patterns of behaviour that determine the commitment to, and the style and proficiency of, an organisation's health and safety programmes.
>
> Organisations with a positive safety culture are characterised by communications founded on mutual trust, by shared perceptions of the importance of safety, and by confidence in the efficacy of preventive measures.'

5.2 Characteristics of organisations with a positive safety culture

A positive safety culture implies that the whole is more than the sum of the parts. The many separate practices interact to give added effect and, in particular, that all the people involved share similar perceptions and adopt the same positive attitudes to safety: a collective commitment.

The synergy of a positive safety culture is mirrored by the negative synergy of organisations with a poor safety culture. Here the commitment to safety of some individuals is strangled by the cynicism of others. Here the whole is less than the sum of the parts. This is evident in organisations where a strong commitment to safety resides only in the safety department.

The CBI (18) has reported the results of a survey of 'how companies manage health and safety'. The idea of the culture of an organisation was incorporated in the report's title Developing a Safety Culture. The dominant themes to emerge were:

(a) the crucial importance of leadership and the commitment of the chief executive;
(b) the executive safety role of line management;
(c) involvement of all employees;
(d) openness of communication; and
(e) demonstration of care and concern for all those affected by the business.

The objective of these and related organisational features is to cultivate a coherent set of perceptions and attitudes that accurately reflect the risks involved and which give high priority to safety as an integral part of shop floor and managerial performance. What is critical is not so much the apparent quality and comprehensiveness of health and safety policy and procedures. What matters is the perception of staff at all levels of their necessity and their effectiveness.

A constant theme of the discussion of safety culture is that it a sub-set of, or at least profoundly influenced by, the overall culture of an organisation. It follows that the safety performance of organisations is greatly influenced by aspects of management that have traditionally not been 'part of safety'. This view has been supported by an extensive research programme carried out by the US Nuclear Regulatory Commission (3). The expert judgement of the researchers who conducted the work believe that the key predictive indicators of safety performance in the US nuclear industry are, in rank order:

(a) effective communication, leading to commonly understood goals, and means to achieve the goals, at all levels in the organisation;
(b) good organisational learning, where organisations are tuned to identify, and respond to incremental change;
(c) organisational focus; simply the attention devoted by the organisation to workplace safety and health;
(d) external factors, including the financial health of the parent organisation, or simply the economic climate within which the company is working, and the impact of regulatory bodies.

The point about these factors is that both (a) and (b) do not concern safety directly; they relate to all aspects of a company's culture. It follows that to make managers manage safety better it is necessary to make them better managers.

The conclusion drawn from the NUREG work and other studies, putting to one side the impact of external pressures and constraints, is that safety depends as much on organisational culture generally as on visible safety management activity. The best health and safety standards can arguably only be achieved by a programme which has a scope well beyond the traditional pattern of safety management functions.

6 AN OUTLINE ACTION PLAN TO PROMOTE A POSITIVE SAFETY CULTURE

There are two barriers which might impede progress. Firstly, the advocates of a positive culture have proposed the simultaneous adoption of every conceivable measure which might lead to improvements. Secondly, the breadth of the concept may make the task of managing improvements appear both abstract and daunting. It must be emphasised therefore that while the outcome of well-conceived plans to improve the safety culture of an organisation may be revolutionary, the plans themselves should be evolutionary. A step-by-step approach is essential.

The first step should be to review the existing safety culture. An action plan may then be prepared on the basis of the findings of the initial review.

Subsequent steps should be driven by analysis of the outcomes of each discrete stage. Some sections of proprietary safety audit systems may help in this evaluation.

Organising for Safety (2) contains a detailed safety culture *Promptlist*. The list, despite being drawn up in the context of nuclear safety, is relevant to safety culture evaluation in any employment setting. It is reproduced here in Appendix 1 in a form where all explicit references to the nuclear industry and nuclear safety have been eliminated.

The distinctive assumption of the Promptlist is that the organisation reviewing its culture already possesses an apparently impressive battery of safety and operating procedures and well-trained staff. The Promptlist rather seeks to establish the adequacy of the steps that the organisation is taking to ensure that everyone in the organisation is genuinely committed to the successful implementation of the safety programme.

The foundations of the Promptlist are:

(a) contemporary models of accident causation;
(b) the key functions of safety management;
(c) the working definition of safety culture (see above);
(d) the published research evidence described and reviewed in detail in (2).

Action plans to improve safety culture should aim to establish an organisation where all employees agree *through a system of communications based on mutual trust* that the organisation's safety procedures:

- are founded on shared perceptions of hazards and risks;
- are necessary and workable;
- will succeed in preventing accidents;
- have been prepared following a consultation process with the participation of all employees;
- are subject to continuous review, involving all personnel.

7 CONCLUDING REMARKS

The shortcomings of traditional safety management were that it was driven by efforts directed at preventing accidents that had already occurred, and by legal duties that gave little discretion to employers (and employees) to target safety effort in the areas they perceived as most appropriate. The new regulations generally encourage a more proactive approach based on management methods that are consistent with good management practice generally, and the transfer of ownership of safety to every employee.

Employers should recognise that a key determinant of successful safety management is the promotion of a positive safety culture, and that good safety performance is not just a matter of the preparation of well-structured company safety procedures.

What is crucial are the attitudes and beliefs of directors and employees at all levels about the procedures. Where everyone believes that the control measures are appropriate in relation to the risk, workable and effective there is the prospect of willing compliance with the safety measures.

The Safety Culture Promptlist presented in Appendix 1 provides a list of questions that employers may wish to consider when seeking to review, and improve, their safety culture.

REFERENCES

(1) HEALTH AND SAFETY EXECUTIVE. Successful health and safety management. HS(G)65. 1991 (HMSO).

(2) HEALTH AND SAFETY COMMISSION. ACSNI Human Factors Study Group third report: organising for safety. 1993 (HMSO).

(3) RYAN, T.G. Organisational factors regulatory research briefing to the ACSNI Study Group on Human Factors. 1991 (unpublished).

(4) DEPARTMENT OF TRANSPORT. Herald of Free Enterprise formal report. 1987 (HMSO).

(5) DEPARTMENT OF TRANSPORT. Investigation into the King's Cross underground fire. 1988 (HMSO).

(6) DEPARTMENT OF TRANSPORT. Investigation into the Clapham Junction railway accident. 1989 (HMSO).

(7) DEPARTMENT OF ENERGY. The public inquiry into the Piper Alpha disaster. 1990 (HMSO).

(8) BOOTH, R.T. Monitoring health and safety performance - an overview. *Journal of Health and Safety*, 1993, $\underline{9}$, 5-16.

(9) HEINRICH, H.W. Industrial accident prevention. 1969, 4th Edn. (McGraw Hill).

(10) REASON, J.T. A framework for classifying errors in New technology and human error. 1987 (Wiley).

(11) ROBENS. Safety and health at work. Report of the committee. 1972, Cmnd 5034 (HMSO).

(12) RASMUSSEN, J. Reasons, causes and human error in New technology and human error. 1987 (Wiley).

(13) HEALTH AND SAFETY COMMISSION. ACSNI Human Factors Study Group second report: Human reliability assessment: a critical overview. 1991 (HMSO).

(14) HALE, A.R. AND GLENDON, A.I. Individual behaviour in the control of danger. 1987 (Elsevier).

(15) ZOHAR, D. Safety climate in industrial organisations: theoretical and applied implications. *Journal of Applied Psychology*, 1980, $\underline{65}$ (1), 96-102.

(16) INTERNATIONAL NUCLEAR SAFETY ADVISORY GROUP. Basic safety principles for nuclear power plants. *Safety Series*, 1988, 75-INSAG-3, (International Atomic Energy Agency, Vienna).

(17) INTERNATIONAL NUCLEAR SAFETY ADVISORY GROUP. Safety culture. *Safety Series*, 1991, 75-INSAG-4 (IAEA, Vienna).

(18) CBI. Developing a safety culture. 1990 (Confederation of British Industry, London).

APPENDIX 1

SAFETY CULTURE PROMPTLIST

1 REVIEW OF ORGANISATIONAL CULTURE - EMPLOYERS

Has the organisation evidence to demonstrate that:

1.1 Communications at all levels are founded on mutual trust?

1.2 All personnel understand, and agree with, corporate goals and the subordinate goals of their work group?

1.3 All personnel understand, and agree with, the means adopted to achieve corporate and work group goals?

1.4 The work practices of the organisation are under continuous review to ensure timely responses to changes in the internal or external environment?

1.5 Managers and supervisors demonstrate care and concern for everyone affected by the business?

1.6 Managers and supervisors take an interest in the personal, as well as the work, problems of their subordinates?

1.7 Managers and supervisors have been trained in leadership skills, and adopt a democratic and not an authoritarian leadership style?

1.8 Workforce participation in decision-making is not confined to peripheral issues?

1.9 Job satisfaction is maintained by, for example, verbal praise from supervisors and peers, equitable systems of promotion, minimisation of lay-offs, and the maintenance of a clean and comfortable working environment?

1.10 The organisation expects the highest standards of competence and commitment of all its employees, but retribution and blame are not seen as the purpose of investigations when things go wrong?

1.11 An appropriate distribution of both young, and more experienced socially mature employees is maintained in the workforce?

1.12 The organisation only recruits suitable personnel, but no automatic presumption is made that individuals are immediately competent to carry out the tasks assigned to them?

2 REVIEW OF SAFETY CULTURE - EMPLOYERS

2.1 Policy, Planning, Organisation and Communication

Has the organisation evidence to demonstrate that:

2.1.1 The Chief Executive takes a personal and informed interest in safety?

2.1.2 The Chief Executive and the Board take explicit and continuing steps to ensure that their interest in, and commitment to, safety is known to all personnel?

2.1.3 A positive commitment to safety is visible throughout the management chain?

2.1.4 Safety is managed in a similar way to other aspects of the business, and is as much the responsibility of line management as any other function?

2.1.5 Safety practitioners have high professional status within the organisation with direct access to the Chief Executive or other appropriate Board member?

2.1.6 Safety committees have high status in the organisation, operate proactively, and publicise their work throughout the organisation?

2.1.7 Managers at all levels, and supervisors, spend time on the 'shop floor' discussing safety matters, and that steps are taken to ensure that all personnel hear of the visits and the matters discussed?

2.1.8 Managers and supervisors spend time commending safe behaviour as well as expressing concern if safety procedures are not being observed?

2.1.9 There are multiple channels for two-way communication on safety matters, including both formal and informal modes?

2.1.10 Safety representatives play a valued part in promoting a positive safety culture, and in particular contribute to the development of open communications?

2.1.11 Specially-convened discussion/focus groups are established to consider the safety aspects of new projects?

2.1.12 Everyone in the organisation talks about safety as a natural part of every-day conversation?

2.1.13 Everyone in the organisation recognises the futility of mere exhortation to think and act safely as a means of promoting good safety performance?

2.2 Hazard Management

Latent (decision) failures: has the organisation taken explicit steps to prevent and detect:

2.2.1 Cases where managers with responsibility for the development or implementation of safe operating procedures fail to:
   (a) search for, and identify, all relevant hazards?
   (b) assess risks accurately?
   (c) select workable and effective control solutions?
   (d) adopt appropriate methods to monitor and review the adequacy of the procedures?
   (e) determine whether foreseeable active failures are likely to be the result of errors at the skill-, or rule-, or knowledge-based levels, or the result of violations?
   (f) minimise or eliminate sources of conflict between production and safety?
   (g) ensure that all relevant personnel have had an opportunity to comment on the procedures before finalisation or implementation?
   (h) ensure that all personnel are adequately trained, instructed and motivated to follow safe operating procedures?

2.2.2 Cases where managers *personally* commit violations of safety procedures or professional good practice?

Active failures: has the organisation taken explicit steps to prevent and detect:

2.2.3 Personnel failing (as a consequence of errors and/or violations) to:
   (a) search for and identify all relevant hazards?
   (b) match their perception of risks to the actual risk magnitudes?
   (c) accept personal responsibility for action?
   (d) follow systems of work where specified, or otherwise adopt a safe method of work?
   (e) continuously monitor and review the magnitude of risks to which they are exposed, and the effectiveness of the steps taken to keep the dangers under control?

Do the organisation's plans for preventing and detecting latent and active failures take explicit account of the following:

2.2.4 Managers, supervisors, and other personnel may tend to underestimate the magnitude of risks:
   (a) with no significant potential (when dealing with major hazards)?
   (b) where the consequences are delayed (for example, a long latent period between exposure and harm)?
   (c) affecting people outside the immediate work group?
   (d) where perceptions may not be adjusted sufficiently in the light of new information?
   (e) where snap judgements are made on the basis of extrapolated information about other hazards?

2.2.5 Managers, supervisors, and other personnel may tend to overestimate their ability to assess and control risks:
   (a) where the hazards have been encountered for long periods without apparent adverse affect?
   (b) where the hazards present opportunities for ego enhancement (for example, public displays of daring (macho image); managers seeking to portray decisiveness)?
   (c) where substantial benefits accrue?
   (d) when the assessment is made by a volunteer?

2.2.6 Managers, supervisors, and other personnel may tend to have an impaired ability to cope with risks:
   (a) when affected by life-event stressors (for example, bereavement, divorce)?
   (b) when under stress as a result of a lack of confidence in the established procedures?
   (c) when they believe that they have no power to influence their own destiny or that of others (fatalism)?

Has the organisation adopted the following measures for improving people's perceptions of risks, and/or ability and commitment to control risks:

2.2.7 A scheme to identify managers, supervisors and other personnel who may:
   (a) be subject to life-event stressors?
   (b) lack confidence in the effectiveness of prevention?
   (c) harbour resentment or distrust of the organisation?
   (d) have an adventurous outlook on risks?

2.2.8 Steps to increase individual belief in their own ability to control events?

2.2.9 Steps to erode peer approval of risk taking?

2.2.10 Discussion groups to talk through individual perceptions of risks and preventive measures?

2.2.11 Safety training founded on:
- (a) a clear recognition and understanding of the likely distortions of peoples' perceptions of risk magnitudes and corrective measures?
- (b) the need for refresher training to counter peoples' changes in perceptions over time?
- (c) feedback of accident/near miss data?
- (d) explanations of not just how a job must be done, but why it must be done that way?
- (e) the need for team building?

2.3 Monitoring and Review

2.3.1 Has the organisation taken explicit steps to determine how its corporate goals compare with those of the local community and society at large?

2.3.2 Is the Board seen to receive regular safety reports, to review safety performance periodically, and to publicise the action taken?

2.3.3 Has the organisation:
- (a) a plan to review, and where necessary, improve its safety culture?
- (b) devised methods for selecting, quantifying and measuring (auditing) key indicators of safety culture?
- (c) reviewed, and where necessary changed, its organisational structure to make manifest its commitment to safety?
- (d) taken steps to ensure safety decisions are acted upon without delay?

2.3.4 Have members of the organisation been trained to:
- (a) carry out a review of safety culture?
- (b) devise and validate key indicators of safety culture?
- (c) prioritise safety culture goals arising from a review?
- (d) draw up an action plan to improve the safety culture of the organisation in priority areas?
- (e) monitor the implementation and effectiveness of plans to improve the safety culture?

2.3.5 Has the organisation made arrangements to encourage reflection on, and elicit the views of all personnel about:
- (a) the overall organisational culture?
- (b) the safety culture of the organisation?
- (c) their perceptions of the attitudes of others in the organisation about safety?
- (d) their perceptions of risk?
- (e) their perceptions of the effectiveness of preventive measures?
- (f) themselves (self-assessment)?

2.3.6 Has the organisation introduced incident investigation procedures which take full account of:
- (a) multi-causality?
- (b) the need to explore the incidence of latent as well as active failures?
- (c) the need to continue the investigation, even when an apparent cause has been found, to determine further causal factors?
- (d) the importance of accepting that the ultimate responsibility lies with the organisation, rather than merely assigning blame to individuals?

3 REGULATORS & SAFETY CULTURE

Has the regulatory body:

3.1 Considered both the positive and potentially negative effects of regulatory intervention in the promotion of a positive safety culture?

3.2 Recognised that a positive safety culture demands that employers should have a sense of 'ownership' of safety?

3.3 Appreciated that key indicators of safety culture go beyond the conventional measures of safety performance?

3.4 Taken steps to ensure that employers' staff are adequately trained to carry out an objective review of their safety culture?

3.5 Taken steps to ensure that employers' staff are adequately trained in the skills necessary to develop an action plan for the promotion of a positive safety culture?

3.6 Recognised that employers' possible misperceptions of the quality of their safety culture may be shared by the regulator's own staff when the regulators have similar backgrounds and are part of the same social system?

3.7 Reviewed, in the context of safety culture, the qualifications experience, and training of their own staff?

Managing professionally

A OSBORNE, BSc, FIOSH and **D BROWN**, BSc, MSc, FIES, FIOSH
London Underground Limited, UK

ABSTRACT

Safety and health at work has been recognised as a necessary management function since the beginning of the century. Over the last twenty years a 'safety language' has evolved which has to some extent taken safety and health from general management into a specialist function. Increasing regulation is accelerating this trend and a concern is that different specialist professions are claiming ownership and propagating different 'dialects' of this safety language. Even the most proactive 'line manager' must be reeling from the different professional messages and language in current usage. The key objective must be to manage our businesses professionally by drawing all relevant knowledge, skills and competencies together. A recurring theme in this paper is that it is neither sensible nor efficient to attempt to subdivide management for the convenience of the professions. The engineering, safety, quality, risk management, personnel, finance etc professions are in danger, by working in professional isolation, of failing to recognise the value of a united, team approach, aimed at managing professionally.

This paper is concerned with recognising and breaking down the professional barriers we have created and seeking out new opportunities to improve *performance*. Costs of poor performance are examined and a way forward proposed to focus all professions on the common management objective of performance improvement.

1. MANAGEMENT SYSTEMS APPROACH

Lord Robens (1972) proposed, that to reduce the incidence of injuries and ill health, the introduction of a 'systems' approach was necessary. Systems essentially consist of standards, prodcures and monitoring arrangements aimed at promoting the health and safety of people at work and to protect the public from work activities. This recognised that technological and engineering improvements alone would not achieve the improvements sought.

Recent disaster inquiries have emphasised the need for Safety Management Systems. In response safety professionals have been engaged upon introducing management systems. The key objectives of a pro-active safety management system are to control the incidence of injuries, meet moral and legislative requirements, and to control loss (minimise waste).

These objectives will vary depending upon the profession and a number of management systems are now in evidence - quality, engineering, environmental, risk, finance etc management systems. Such a separatist approach has introduced much duplication, inefficiency and, worse still, confusion for those who must apply the multitude of management systems. Such systems are either developed 'in house' or 'off the shelf' systems, frequently driven by proprietary assessment systems known to the different professions.

Proactive safety management systems integrate many of the traditional areas to eliminate injury and ill health at work and contain elements such as, leadership, policy, objectives, organisation, communications and monitoring in common with other management systems.

2 TOTAL QUALITY MANAGEMENT

Total Quality Management (TQM) has been advocated as a process which if adopted is capable of solving any safety problems. Sir Anthony Hidden the Chairman of the Clapham Junction train collision Inquiry, whilst supporting and recognising the value of TQM is sceptical that it is the panacea.

'TQM was not specifically designed as a safety initiative, but with the implementation of improving standards, procedures and training there would be safety benefits. If fully implemented TQM will eventually form an important tool ----- in ensuring safety -----'.

'However ----- the intended implementation of TQM cannot be a substitution for good management, safe working practices, -----'

Hidden recognised that TQM is primarily driven from a product and customer satisfaction stance. Whilst safety and service environment can be easily incorporated within TQM the need to incorporate all business risks, is not universally recognised nor rarely applied.

3 INTEGRATED APPROACH

Safety professionals have, for decades, advocated that safety cannot be managed in isolation and, for example, safety training should be an integral and essential element of all training courses, not, dealt with separately. This is, in part recognised by the Health and Safety Executive (HSE) in their publication 'Successful Health & Safety Management' which quotes a number of companies who have a more integrated approach to safety. The publication goes on to discuss the synergy between safety and quality.

The challenge for the future is for each profession to recognise that they share a common objective, of improving management and business *performance*. The establishment of a common management framework with emphasis on human factors would be a great prize for all professions (see section 7).

4. PERFORMANCE CULTURE

In recent years, and again arising from public inquiries into major disasters such as the King's Cross Fire, the capsize of the Herald of Free Enterprise, the Piper Alpha Oil Rig fire and the Clapham Junction train crash, considerable emphasis has been placed on safety management systems and the creation of associated safety cultures. In parallel Quality professionals, in response to BS5750, are promoting the need for Quality Management Systems and a quality culture. The introduction of the Environmental Protection Act & BS7750 are leading to the need for an environmental culture. Furthermore even Quantified Risk Assessment cultures are being demanded by the professionals! Those poor Line Managers!!

Organisational culture is shaped by the prevailing management style and how teams and individuals are focused upon clear objectives, and a shared purpose. Furthermore, it is the extent to which the whole workforce are recognsied and their views harnessed. Each profession (particularly Safety and Quality) is unnecessarily defining its part of 'culture' which 'belongs' to their part of the discipline of 'management'. An organisation's culture is in reality indivisible and uniquely integrated. No useful purpose is gained by attempting to create divisions. To the contrary, it is counterproductive, inefficient and confusing to the proactive manager.

An organisation's culture can be positively influenced by providing the opportunity for the workforce to determine how standards will be met and performance monitored. It has been demonstrated that a commitment to value individuals and teams <u>will</u> improve all performance; quality, production, service, efficiency safety etc. Professional management requires a balance to be achieved between sometimes (though not often) conflicting elements of performance. A key attribute of the modern professional manager is to be able to manage co-prime objectives. Risks have to be identified and managed. A positive organisational culture will support and encourage the management of risk at the 'point of control', by the people most able to do so. Culture has profound implications on such important issues as the first line supervisor always erring on the side of safety when balancing against production issues.

It is essential that the structure of an organisation enables workforce participation in the making of procedures and rules. The workforce have the relevant knowledge and information in this area with the professionals (whoever they are!) simply steering and enabling. Lord Robens and Cullen supported workforce participation 'the workforce must be encouraged to participate fully in the management making and monitoring of the organisation of safety' (Robens).

The authors believe that the prescriptive nature of legislation (for example the Safety Representatives and Safety Committees Regulations 1977) in this area is now holding back further improvement because it is too narrowly focused (see Section 5).

Performance improvement of production, customer service, quality or safety requires leadership and good management (see Section 7). Furthermore involving the workforce in the design and implementation of performance improvement programmes will have a very positive influence on organisational culture.

> 'Concern for safety in relation to the product of any industry must by necessity be driven by a number of factors : some will be humanitarian, some commercial and some a combination of both' (Anthony Hidden - Clapham Inquiry)

R H Ramsay (Industrial Accident Prevention Assessment) summarised these sentiments 'I cannot stress too strongly that production and quality management are inextricably inter-related with workplace safety and in particular, loss control in general.'

Emphasis must be placed upon developing the organisational culture to achieve a balanced, improved performance. The promotion of separate cultures by each profession is counter productive and encourages a partisan attitude of the professions.

5. LEGISLATION

The role of safety regulatory systems, whether external or internal, is to promote company endeavours to improve safety performance. In 1970, Lord Robens found nine main groups of statutes supported by nearly 500 subordinate, prescriptive statutory instruments.

The HSE's Director General recognised the need for further revision of the regulatory system and has stated the objective: 'to reduce law that is fussy or demands unnecessary detail'. This is quite a challenge in view of:

1. the plethora of Directives emanating from the European Community

2. pressures not to dilute existing regulations due to 1 above.

3. pressures for even more regulations from some bodies.

The privatisation of BR is having an impact on point 3 above.

However critical we may be of safety legislation, it is not uncommon for legislation to be used as a smoke screen by both safety professionals and others.

'safety management is different - there are legal requirements'. This is often claimed by safety professionals and yet it does not drive real improvement at the front line.

Business Cases for project expenditure are often made on the basis of statutory obligations. This has two problems. Firstly compliance with the law does not necessarily reduce accident rates and secondly, compliance efforts can be 'overegged' to take account of that particular project the manager has always wanted to implement. Both of these problems can have the effect of not addressing priority safety risks because money and management resource has been used up on lower priority items. Misuse of the safety argument is rife by managers, safety professionals, trade unions and the workforce.

The Health & Safety Executive are accountable for estimating the cost benefits of new legislation. Managers from all disciplines must also be accountable for establishing the cost benefits of implementing legislation.

Lord Robens had a clear vision of the respective roles of legislators and management:

'too much reliance is placed on state regulation';
'The primary responsibility lies with those who create and work with risks.'

Robens and the Chairmen of Public Inquiries have found the state regulations ineffective in preventing major disasters Even the companies practising self-regulation continue to require the 'protection' of prescriptive legislation ie the checklist approach. If the aeroplane had an airworthiness certificate and the pilot was trained to current standards when the air disaster occurs then improvements will not be made. This approach does not engender a continuous improvement spirit for safety.

Self- regulation, as recommended by the Robens Committee has not been a resounding success. There are many reasons for this but we believe the principal ones are:

i) the inadequacy of the regulatory system; prescriptive rather than goal setting regulation; inquiries focussing on the proximate causes not the underlying organisational and management shortcomings; enforcement officers, in response to the legislative requirements, adopting too prescriptive an approach. The latter discourages employers and workforce accepting the responsibility for and ownership of safety;

ii) apathy has continued amongst many employers and much of the workforce; workforce participation (recommended by Robens) has only been lukewarm. Safety Representatives have not been effective - partly due to Trade Union preoccupation with other matters and partly due to employers failure to recognise their full potential;

iii) although it has been recognised that the management of safety and health contributes to the creation of a pro-active organisational culture and achievement of broader business objectives, action has been slow. Leadership skills are recognised as a key to achieving business objectives but safety performance has not been high on the agenda.

iv) perceived competing interests; production, quality, customer service seem in competition for priorities and resources;

v) compartmentalisation of the professions, and 'turf protection', have not been conducive to developing an integrated management approach which harnesses the full energies and potential of the workforce.

The opportunity to improve safety performance through those in the best position to do so - the workforce - has not been lost. Organisational behaviours are changing positively. *'In the last decade in particular, there have been significant improvements in the development of skills and educational standards of the workforce, which means the need to change relationships and attitudes towards employees ...'. 'Employees want to be part of the group ... they want to be to a greater extent now in control of their own destiny'*. (S N Malcolm Field, W H Smith).

The introduction of the Management of Health and Safety Regulations, together with the repeal of prescriptive legislation following the HSE review and 'downsizing' of regulations coupled with the contemporary focus on business objectives must be exploited fully in the evolvement of self-regulation.

6 COSTS OF POOR PERFORMANCE

'The promotion of safety and health is ... a normal management function - just as production or marketing is a normal function' (Robens). There are costs and benefits associated with the achievement of improved safety performance. The cost/benefit ratio is measured in terms of total cost of mitigating against the loss, balanced against the loss. This is the basis of the concept of 'reasonably practicable' but application is complex since the resources utilised in mitigation will have benefits in other activities - an improved access route may result in both improved safety and production, similarly any loss may involve safety, quality, production, property, finance etc. The by-products of safety improvement can be very worthwhile and it is important that wherever possible, solutions are explored which give an optimum return across safety, quality and efficiency.

The HSE Accident Prevention Advisory Unit (APAU) recently undertook research into 'avoidable accident costs' in five diverse companies. APAU summarised these losses as:

- one organisation 37% of annualised profit
- another organisation 8.5% of tender price
- a third organisation 5% of running costs

All very significant avoidable 'overheads' eating into profit margins. Add the quality failures and the prize of getting it right is enormous and yet, the professions are not tackling these issues in any kind of coordinated or concerted way.

At London Underground, in conjunction with William Lynn Associates, we engaged in the measurement of poor performance in safety and quality activities. The analysis was carried out on one Line on the Underground. The power of the approach is that it is capable of providing comparable analysis, for benchmarking purposes with other lines, railways and companies. The methodology involved the following techniques:

- statistical analysis of performance information
- compliance audits of management systems
- structured observation of people at work

The observation phase involved two thirds of the front line staff. In the course of the project over two hundred examples of non-conformances (a proportion of which were safety related) were identified. Each was fully documented and validated by line management.

Individual costings, based upon salaries, cost of train service delays and the cost of delay or stoppage to other key assets - eg lifts and escalators, were allocated to each non-conformance. Assessments were made of opportunity costs, rescheduled production costs, scrap, wastage and repair. No prejudgement was made as to the categories of non-conformance to be used, instead these were post event categorised to form a costs hierarchy.

Using the data collected and the cost allocations it was possible to determine the cost/ benefit ratio and design a 30 week programme of correction actions. In the case of the line studied, the 'Cost of Poor Performance ' was estimated at £7.6m and a profit/recovery goal of £2.5m set. This represents 7% of the total line running cost which when realised will be available for reinvestment, to further improve performance.

This process has identified where management systems are not working, controls not in place and opportunities for improvement.

We believe that this approach is an essential starting point for a performance improvement programme. It is measurable, targetted, ensures the priority areas are tackled first, and, it capitalises on a unit of measurement that all professions relate to- namely money. The approach has safety as an integral part.

7. THE WAY FORWARD

This paper has set out to show that neither traditional safety management systems, nor legislation alone are sufficient in themselves to achieve the required improvements in safety performance. It is essential that amendments are made to the regulatory and enforcement systems to ensure that more effort is directed at organisational culture. Lord Cullen found that despite regular management meetings at which safety was discussed; 'near misses' reporting systems; specialist safety advice; safety committees; review and audits - *'all the semblance of a well managed safety conscious organisation - its reality was that safety was not being effectively managed'*. The reigning organisational culture, attitudes, management and workforce were not delivering the required performance levels.

The Robens Committee advocated that managers should provide clear objectives and a framework in which the workforce can operate to meet the objectives. Modern management theory advocates that managers provide leadership and coaching to enable the workforce to take the authority and responsibility to manage workplace risks. Managers must create a climate in which the workforce is valued, respected and developed. Clear, effectively resourced, policies must be established which are focused upon delivering business results.

The key areas of leadership, people and resource management, policy and strategy, employee and customer satisfaction, are identified and incorporated into The European Quality Award (EQA) framework (see Appendix 1), introduced by the European Foundation for Quality Management (EFQM). The essential difference in the EFQM approach from traditional safety management approaches is that it is not prescriptive but provides a framework for improvement. Safety is an essential component of the framework but the real benefits will accrue from using the entire framework as a safety improvement model. All efforts are concentrated upon achieving results (performance) through harnessing all the potential energy, skills and competencies of the workforce in processes which meet the needs of customers and society. It goes further by requiring that the workforce must have the authority, knowledge, skills, competencies and information to deliver improved performance. This, if adopted, would give true realisation to self-regulation. The linkages between organisational culture and effective self-regulation are critically important.

The EQA framework requires that the best management techniques and tools available are utilised. Approaches such as this must be seen as the way forward if we are to improve safety performance alongside and instep with other performance improvements. The fact that the framework has been inspired by quality professionals should not put other professionals off. The framework can be applied to any improvement opportunity including environmental issues.

8. SUMMARY

We set out in this paper to challenge the approaches which have been taken to introduce safety management and audit systems into companies as a means of improving safety performance. In doing so we have touched on some of the organisational and human behaviour factors which are key to real and lasting performance improvement.

We have asserted that prescriptive legislation is counter-productive to the effective management of safety (and hence self-regulation). The concept of self regulation we suggest, has not progressed but the opportunity is not quite lost.

As engineering and safety professionals at the Management of Safety Conference, we must integrate the effort companies are putting into their separate programmes for safety, quality and efficiency into an overall performance improvement. We must all work to a common framework, perhaps the EQA framework?

9 REFERENCES

Safety and Health at Work
Volume 1 Report of the Committee 1970-72
Volume 2 Selected Written Evidence
Chairman Lord Robens
Command 5034
Her Majesty's Stationery Office (HMSO)

Department of Transport
Investigation into the King's Cross Underground Fire
Desmond Fennell OBE QC
Command 499
October 1988
HMSO

Department of Transport
Investigation into the Clapham Common Railway Accident
Anthony Hidden QC
Command 820
November 1989
HMSO

Department of Energy
The Public Enquiry into the Piper Alpha Disaster
The Hon Lord Cullen
Command 1310
November 1990
HMSO

The Health and Safety Executive
Successful Health and Safety Management
Ref HS(G)65
1993
HMSO

The Health and Safety Executive
The Cost of Accidents at Work
Ref HS(G)96
1993
HMSO

BS5750 British Standard Quality Systems
Parts 0-3 1987
Part 4 1990
British Standards Institutuon, 2 Park Street, London

European Quality Award
European Foundation of Quality Management, B-1200 Brussels

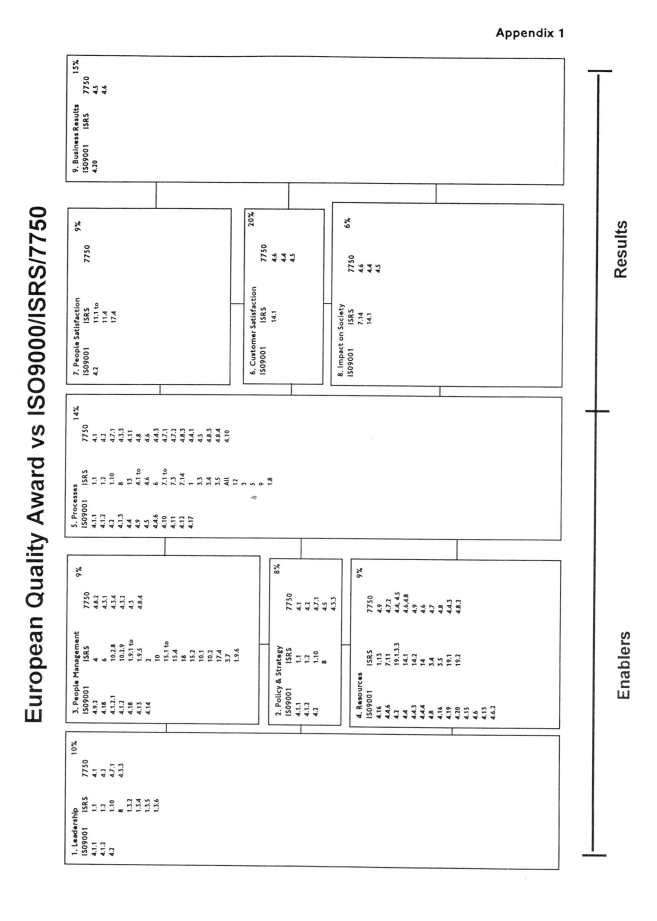

Appendix 1

Company strategy for the management of safety

F T DUGGAN, BSc, CEng, FICE, FIHT
Tarmac Construction Limited, Wolverhampton, UK

We are all aware of the great disasters of the last decade; Chernobyl, Piper Alpha, Kings Cross, Zeebrugge, Clapham, Lockerbie and Hillsborough, but beneath these major disasters lies a catalogue of other tragic accidents and deaths which do not make the headlines.

In the construction industry alone there are 100 - 150 fatal accidents every year, 2 or 3 every working week.

During my 30 years of working in the construction industry, over 4000 construction workers in this country, have lost their lives, over half a million have been very seriously injured and millions have suffered injures while carrying out their work.

All involved in construction have learned many lessons and those who are in positions of influence must ensure that these lessons are not forgotten when we come to developing a strategy for management.

Before I turn to look at a more detailed company strategy for safety, I would like to look at the reasons why the approach made to safety in the past has not been a success. Sometimes it can be said that looking back serves very little purpose, but I think it is also wise to reflect on the lessons we can learn from the past and build on these for the future.

HISTORY - WHY HAS LEGISLATION NOT SUCCEEDED?

The history of protective occupational health and safety legislation can be traced back to the passing of the first factory statute in 1802 as a result of concern about the appalling working conditions of children in the cotton mills. The next 170 years saw the piecemeal extention of protective legislation to cover other types of worker, workplaces and hazards. The ad hoc fashion in which the structuring framework for occupational health and safety evolved meant that a highly complex body of law had developed by the early 1970's.

Managers were being restrained by legislation. Not surprisingly therefore, in 1972 the Robens Report concluded, "first and perhaps most fundamental defect of the statutory system, is simply that there is too much law." The Report accepted the argument that the sheer mass of law was counter productive. People were conditioned to think of Safety and Health at work as in the first instance, a matter of detailed rules imposed by external agencies. Legislation of itself therefore, has not succeeded, a fundamental shift in our approach is required.

WHERE SHOULD COMPANIES BEGIN?

The earlier those of us in management positions begin to recognise our responsibilities, the better. The experiences that our young people gain early in their careers are seldom forgotten. All of us are at our most receptive when setting out to learn our trade.

Young people of to-day do not have the opportunities to develop, what I would call an awareness of danger, where in the past this could be developed before setting foot onto a site. We had the opportunity to gain awareness either by National Service, some came from agricultural backgrounds, many were slowly weaned into professions by apprenticeships and all of this gave us the breathing space required before accepting responsibilities for others.

The opportunity to work alongside tradesmen and ex-servicemen with years of experience enabled us to learn in a slow and methodical manner. We have now quickly moved into a technological age and the rate of change means that the opportunities for injuries at work increase with change and with constant regularity. The overall rate of accidents has failed to decrease over the last decade and in fact the number of reported injuries has risen considerably.

The opportunities to learn the trades have disappeared, and yet we rely on our young people to accept responsibility very early in their careers. Companies must play their part in developing safety awareness, not only in the workplace, but also in our Schools and Colleges.

HIGHER EDUCATION

Let us turn to Higher Education to see what influence we can have before our young people start work in any of our businesses.

Safety is not a particularly suitable subject to teach on a theoretical basis, as most students have no site experience and many safety principles are difficult to appreciate as they do not have a direct relevance when talked about in a lecture room. Students are less concerned about learning safety rules and regulations, than learning material which will be useful to them in passing their examinations. Higher education establishments are academic institutions, teaching subjects in quite a pure form. Many other more traditional subjects must compete for the limited lecture time available and it is easy to neglect the teaching of safety.

Many Universities now take advice from industry when developing their courses and contractors, business people and others must continue to use whatever influence they have to stress the importance of increased safety training.

Higher education does not require any specific length of time to be spent on safety. If particular requirements for safety lectures, course work and questions in, say construction management or other management examinations were to be introduced, greater safety awareness could be developed. Many Universities are indeed doing this and this is a start.

By including safety questions in examinations, students might be more likely to learn and to think about safety. Coursework could include research and report writing or perhaps a visit to a factory or site, but with greater importance placed on the potential dangers and hazards of that workplace. The safety techniques of risk assessment could be useful to introduce at this stage.

Realistically, however, we must all accept and recognise that higher education cannot do much more than develop an awareness of the importance of safety. On the other hand they can give an extremely valuable start in kindling the awareness we require.

SAFETY STRATEGY

All the knowledge will be useless if it is not channelled into a suitable plan with aims and objectives.

There are many similarities in producing a safety strategy when comparing this with any other business. The business must know what it is trying to do, set up the organisation to carry out these aims, have the available resources to produce the goods or services, measure its success in achieving the goals and then review its future plans.

Any safety strategy cannot do much worse than follow and develop the general principal as set out in the Health and Safety at Work Act 1974, i.e. to produce a commitment from senior management in the form of a safety policy and to set up a suitable organisation with arrangements for carrying out that policy.

The success of any safety programme depends directly upon the most senior management. In my own company the Chief Executive is on record stating that "safety is my No.1 priority".

This message is accepted and acted upon by all the Managing Directors of the Construction Divisions of the Company.

Safety is an item on the agenda of every Board meeting and our Safety Services Department is given every possible support from the Board.

My company, like many other successful companies, has set up organisations which are committed to achieving high standards of health and safety.

These systems are developed following recognised good principles such as:

a) Health and Safety can contribute to improved business performance by preserving and developing human and physical resources. In practice costs and liabilities are reduced and there is a means of expressing corporate responsibility. This must be led from the top.

b) business leaders developing appropriate organisational structures and a culture which supports risk control and secures the full participation of all members of the organisation.

c) the need to resource and plan policy implementation adequately.

d) the systematic identification and control of risk as the only effective approach to controlling injury and ill health and loss prevention.

e) the need to scrutinise and view performance to learn from experience.

f) interrelating quality, health and safety.

g) the establishment of safe working practices within routine processes.

h) the supply and wearing of protective clothing, arrangements and selection of safety equipment.

i) reference to a separate fire manual and appointment of fire wardens to reduce the risk of fire and explosions.

j) assistance with staff welfare problems.

k) provision of first-aid equipment and first-aiders.

l) using safety advisers and communicating how to seek advice.

STANDARDS TO BE ACHIEVED

These principles then need development into a series of arrangements and standards which are prepared in procedures and safety standards manuals.

These procedures and standards manuals receive wide circulation within the company and are well used for reference so that a consistent quality can be achieved.

However, none of the above can work unless people can carry out their duties competently and in order to become competent suitable and effective training is necessary.

Training must start from the first day at work with induction.

SAFETY INDUCTIONS

It is a sobering thought for all of us that if we just check in a worker, give him a hard hat and tell him to go to work, then it means that we do not care too much for safety. If we reflect on the fact that almost 50% of fatal accidents in construction occur either in the first day or the first week on site, we could be sending him to his death. In our Company all employees, staff managers, young graduates, operatives and sub-contractors receive an induction course. It includes a welcome to the place of work, safety training, introduction to job specifics rather than generalities, and it can in some cases be a slide, tape or a video presentation on our larger contracts.

Points to be included in an induction course must be:

a) all employees should read the Company Handbook.

b) they heed their personal safety and that of colleagues and the public at large. This is a requirement, not only by law, but of their employment by the Company. Breach of that duty can result in a reprimand or in extreme cases, dismissal.

c) how to obtain first-aid.

d) what to do in the event of a fire.

e) what to do in the event of an accident.

f) how to obtain protective clothing and safety equipment.

g) brief outline of the likely hazards employees may encounter together with examples that have occurred.

h) a brief outline on further safety training and information that is required for personnel safety on site.

Younger people, graduates and technicians on site for the first time must be given special consideration, because of their vulnerability. Some of our contemporaries on the Continent ask their young graduates and technicians to spend a short period of time work-shadowing with a more experienced member of staff, to enable them to develop the awareness which is absolutely essential when walking on to site for the first time. I recommend that we adopt this practice?

Again let me remind you that the most likely time for accidents is during their early days in a new and different environment.

ONGOING TRAINING AND EDUCATION IN SAFETY

Ongoing formal training must be part of safety development. Courses should be organised on various aspects of safety.

At regular intervals talks and lectures should be given on various issues and all staff expected to attend. All business units should plan with their Safety Adviser, a programme for appropriate safety training of all their employees.

Such specific courses can pass on a great deal of safety information. The danger however, is that courses can sometimes seem boring or long winded, and if interest in safety has not already been generated, very little will be learned. One-off events do not maintain safety awareness in the day-to-day running of the company, continuous and sustained efforts must be maintained.

Ways to make safety more interesting and to give it a higher profile are:

1. evening 'socials' including safety quizzes with prizes given to the winners. This is a lighthearted yet informative way of improving awareness.

2. monthly safety bulletins in the form of a news sheet. The record of the various sites is shown in a league table, accidents are reported and particular hazards 'spotlighted'.

3. eye catching posters displayed around the site.

4. the site with the best safety record awarded a prize. (extended on some sites to individuals who try to run a safe site and are seen to be concerned for safety are rewarded).

5. the safety video mentioned earlier shown to everyone joining a new site. Although it might have been seen before on previous sites, it serves as a useful reminder about precautions to be taken and maintains safety awareness.

6. specific site induction training covering the hazards of that particular site.

SAFETY AND TOTAL QUALITY MANAGEMENT

A successful company strategy cannot be developed without considering the influence of a Total Quality Management System. Other speakers are addressing this subject, suffice to say at this time that the disciplines imposed by a Total Quality Management system provide a company with an excellent base on which to build a system complete with such vital elements as organising, planning, measuring and auditing.

For far too long people have measured safety negatively. They have measured failures, given lists of faults and have worked to the policeman principle that one should find things wrong and then take action - often punitively.

My company have developed what we believe is a much better system which actually measures success in safety. We firmly believe that prevention is not just better than cure, if we prevent something going wrong we do not have to cure the results.

We strongly believe that positive actions will result in less failures and we therefore, audit positively by giving credit for things done correctly. This system is well liked by all our managers who are recognised for the efforts they are making. They also appreciate what they need to do to gain more recognition and in fact our Safety Award Schemes are now judged on the results of our Positive Safety Measurement Audits.

ACCIDENTS

I commenced my talk by commenting on the number of accidents which have occurred in the industry.

Perhaps the ultimate measure of a company's success is the number of employees who die or are injured whilst at work.

The employee accident frequency rate for Tarmac Construction has fallen considerably over many years.

In 1992 our rate was only 2/3 of that in 1987.

Even that does not tell the whole story. Our accident incident rate is well below the average for the industry as published by HSE and only 30% of the real figure which HSE estimate from the Labour Force Survey.

CONCLUSION

In conclusion therefore, a company which is successful in business can be equally successful in safety by using similar techniques, but success needs commitment from the top, the correct organisation and a constant effort to continually improve.

New technology – the implications for management

V L MAYATT, BSc, PhD, FRSH
Accident Prevention Advisory Unit, Merseyside, UK

SYNOPSIS The Health and Safety Executive (HSE) has looked at the management implications associated with the introduction of new technology. This work has been carried out by HSE's focus of expertise on health and safety management the Accident Prevention Advisory Unit (APAU). The project has involved working with industry to identify examples of new technology, associated management problems and health and safety issues. A range of issues for managers which impinge upon health and safety have been identified: human factors, training, accidents, derivation of standards and communication. Risk assessment is a particular problem. These issues are developed in the paper.

1. WHAT IS NEW TECHNOLOGY?

Technology can be taken to mean the application of science whether within industry or other spheres of life such as medicine. Part of what made Britain great in the Victorian era centred on technological achievements in the fields of civil, construction and mechanical engineering, such as this country's railway system, Brunel's bridges, and the Crystal Palace.

New technology, that is how science is applied now, is extremely diverse in direct reflection of the enormous progress within science over the last century. In industry, examples of new technology include the use of robots to construct cars, automated vehicles to select and store stock in warehouses, chemicals, such as vitamin C, once produced by conventional chemical reaction, are now produced by biotechnological means. Future developments within biotechnology may lead to microbial mining, where it may be possible to release hydrogenous materials from fossil fuels and oil-bearing strata (1).

New technology is not just confined to the workplace, it touches all aspects of life: the use of lasers at pop concerts, sophisticated instruments and developments, such as monoclonal antibodies, used in medical diagnosis and treatment, and the use of advanced pharmaceuticals in health care. The televising of the recent Gulf War dramatically demonstrated the state of the art technology associated with weapons and military intelligence. Food production is increasingly affected by technological developments, for example the use of Baculovirus insecticides as alternatives to chemical insecticides (1).

2. WHY DOES INDUSTRY NEED NEW TECHNOLOGY?

Industry is competitive, to make money and survive it has to manufacture new and better quality products, more cost effectively and efficiently. Inevitably that means harnessing new technology. It has been stated that one of the most important factors in maintaining a country's standard of living is high-technology industry; electronics, aerospace, chemicals and biotechnology are, within the USA, regarded as the drivers of productivity, prosperity and economic growth (2). Therefore to survive and for national economic success, industry drives itself, and is in turn driven, to acquire new technology.

3. MANAGEMENT ISSUES

3.1 Attitudes to New Technology: Human Factors

One of the key issues in relation to new technology is that people respond to it differently and do so for a number of reasons. Within the workplace, employees may question the benefits, they may consider that their jobs will be lost when, for example, robots are introduced to undertake welding tasks they have always done. Employers may foresee the benefits of better or new products, less workspace taken up or faster production schedules. Middle managers may not directly reap the benefits whilst also having to deal with an unenthusiastic workforce.

Luddites may be present both within managerial and shopfloor levels, and those adopting such a stance may do so for different reasons. For example, where a piece of technology is associated with being developed and used within a particular country, views may stem from attitudes to the country concerned rather than the technology itself. From a historical context it has been concluded that when technology is untried and associated with significant financial risk there is more likely to be resistance amongst the workforce (3).

The OECD talks of a "new biotechnology" evolved from more recent developments such as genetic engineering and cell fusion. New developments in these areas will provide a new source of drugs, vaccines and diagnostic reagents, a better understanding of life and health as the genetic code is unravelled. These developments provide tangible benefits to mankind but have profound socio-economic consequences in addition to issues of safety and ethics. Consumers are concerned that food produced by biotechnological means is different from conventionally produced food. There is a fear about the unnaturalness of some developments such as human factor IX production in sheep, and transgenic animals. Some consider this is playing God and manipulating life itself.

It has been concluded that political attitudes and economic costs are inter-related and have, for example, combined to halt the implementation of the nuclear programme in the USA (4). Public awareness of risk from new technology in general can be affected by the success of lobbying in one area of new technology, for example, the nuclear debate has fuelled the public expression of concern about new technology in general.

Attitudes to new technology can be influenced by discussion, debate and education and where common benefits are perceived. An example of this later point is the case of mass production of the Model T Ford in the USA. In 1914 Henry Ford introduced the moving production line. Line workers were paid $5.00 per day when the average industrial wage was $2.50. The 9 hour working day was reduced to 8 hours so that a third shift could be introduced. In this instance the introduction of new technology was associated with social innovations from which both Fords the motor company and its employees profited.

3.2 Risk Assessment

Assessing the risk from new technology is of paramount importance and a particular management challenge. Risk may have to be assessed in the absence of detailed knowledge, case history and practical models on which to base comparisons. Health and safety cannot be compromised. Risk has to be assessed in the light of the best available knowledge and revised as knowledge accumulates and informed opinion changes.

Industries based on new technological developments experience different hazard and risk perceptions in different countries. For example, within the USA nuclear power was originally promised to be cheap and satisfy the continent's large appetite for electricity. Now it is perceived to be expensive because of accidents and their environmental consequences. (Public opinion in America was reputedly against nuclear power prior to the nuclear accident at Three Mile Island in 1979). In France, 75 per cent of electricity is supplied from nuclear power, where it is regarded, as in Belgium, as a necessary evil for a period of time (4).

Whether laboratory or industrial scale biotechnology work is involved, HSE's Advisory Committee on Genetic Manipulation (ACGM) and its predecessor the Genetic Manipulation Advisory Group (GMAG) has assessed risk on a case by case basis. Part of the success of this risk assessment strategy for biotechnology is attributable to consultation between interested and informed parties, risk assessment prior to work getting underway and review of assessments in the light of experience and knowledge. Whilst "ACGM equivalents" may not be merited in other aspects of new technology, there is, for managers, much to commend this method of working, which can be mirrored at a local level and in other technological areas. Multi-disciplinary teams are needed for this task as the knowledge and skills needed to assess risk is unlikely to reside in one individual or individuals of the same discipline or experience.

New technology can, however, reduce risk, for example the decomposition of polychlorinated biphenyls (PCBs) by genetically manipulated micro-organisms (GMOs). Flexible manufacturing systems may remove the need for manual handling and direct contact with dangerous machines so it therefore impinges upon the overall risk equation in a number of ways.

3.3 Accidents and Near Misses

Where understanding and assessment of risk is developing, the investigation of accidents, cases of ill health and near misses plays a key role. Allied to this is the keeping of records of work undertaken, exposures and control measures. In the 1980s there was concern about an apparent cluster of cancers amongst laboratory workers at the Institut Pasteur in Paris. Questions were asked about the nature of the work associated with newly recognised oncogenic viruses. However the statistical validity of the cancers and the more conventional risks from carcinogenic chemicals also handled in the laboratory had also to be investigated.

More recently HSE has announced (5) a re-examination of the health risks to women working in semi-conductor manufacturing. The study follows on from reports in the USA of spontaneous abortion amongst women where work is associated with exposure to ethylene glycol ethers. Information will be sought from the manufacturing sites to identify existing and past female employees. The information sought will be used to show whether any association between work and ill health effects can be found in GB as has been reported within the USA.

Accident triangles are used to portray the relationship between accidents of differing severity. Examination of occupational accidents reveals that the tip of the triangle consists of a small number of accidents involving death and major injury, whilst the wide base of the triangle is made up of a relatively large number of minor accidents or near misses which do not result in injury (6). Whilst lessons can be learnt from the investigation of all accidents to establish immediate and ultimate causes, the investigation of near misses is of relevance to new technology. This is because there will be more of them from which lessons can be learnt, enabling risk and work procedures to be reassessed accordingly.

3.4 Derivation of Standards for Safe Working

The semi-conductor industry claims that part of its success is attributable to effective liaison. There has been co-operation and pooling of knowledge between involved companies has played a key part not just in risk assessment but also in the derivation of standards for safe working. Liaison at joint working group level with industry associations such as the Chemical Industry Association (CIA) and the Semi-Conductor Safety Association (SSA) both within the USA and Europe and HSE working groups, have all played an important part. Debate in such fora about risks to health and safety and the considered derivation of health and safety standards has undoubtedly assisted the managements within this industry to allay concerns for health and safety and to introduce the necessary arrangements to prevent or control risks to health and safety.

3.5 Communication

Communication between managers, between managers and the rest of the workforce is an important part of the successful management of health and safety (7) and is a major factor in relation to new technology.

Communication and the timing of new technology introduction has been a particular issue for Rolls-Royce in its successful introduction of an advanced integrated manufacturing system (AIMS) used to produce turbine and compressor discs for gas turbine engines. AIMS is a flexible manufacturing system consisting of an integrated group of versatile machine and process cells that manufacture different discs, directed by a computerised central control system. Automatically guided vehicles (AGVs), controlled via cables buried in the factory floor, transport components throughout the system.

Before the introduction of AIMS, it took 6 months for a rough machine forging to become a finished item, as over 30 separate precision manufacturing operations were involved. AIMS was designed to reduce this production period to 6 weeks, so that with 24 hour processing over 50 finished disc parts could be produced per week. The spin off for Rolls-Royce was consistently high quality products produced economically and efficiently.

Rolls-Royce recognised at the outset that the introduction of this system would take time. Communication throughout the site was intensive and included the production of scaled models of the new factory layout, explanatory brochures, presentations to the workforce and cardboard replicas of the AGVs. In all, 8 years of planning and consultation with employees took place before AIMS was introduced.

3.6 Recruitment and Training

The semi-conductor industry in which thumb-nail sized silicon chips are produced, which undertake the work for which over 1.5 million transistors would once have been needed, does not regard itself as new, but rather a distillation of many disciplines culled from elsewhere. Chips are produced under extreme conditions of cleanliness where air is HEPA (high efficiency particular air) filtered into the workplace; the fabrication stage is carried out under positive pressure to reduce contamination. Individuals employed in such areas have high standards of personal hygiene and wear complete sets of protective clothing designed to reduce to negligible levels chip contamination. Biological contamination of a chip can result in a short circuit. The pristine space-age appearance of the manufacturing areas therefore appears vastly different from a traditional manufacturing site.

The people attracted to work in this industry tend to be young and intelligent. The manufacturing process is detailed, precise and exacting. Process engineers draw up the implanter process instructions and operators are given 5 to 6 pages of tabled instructions which have to be accurately followed.

The nature of the work therefore dictates the type of individuals attracted to work within the industry; working in full protective clothing under clean room conditions does not suit everyone. Intensive training is provided to ensure that new employees understand the nature of the work and the exacting process stages. The ability to follow detailed documented instructions is paramount together with good manual dexterity.

Large sophisticated organisations are engaged in this work. The financial outlay needed to build a manufacturing plant is prohibitively expensive as production equipment and control mechanisms for both health and safety and product protection are extremely complex. Such organisations are therefore able to invest heavily in training programmes. Training is an integral part of success within the industry as there is no margin for error in the finished chip and the potential health and safety issues particularly from toxic and explosive gases are high.

Training of all employees including managers is important if health and safety is not to be compromised. This is especially true for manufacturing processes based on new technology where unfamiliar techniques and methods of work may be involved. Unless proper training is provided the significance of safe working practices will be lost.

REFERENCES

(1) ADVISORY COUNCIL ON SCIENCE AND TECHNOLOGY. Developments in biotechnology. 1990. HMSO.

(2) ROSS, I. M. Sun Sets Over Silicon Valley. New Scientist, 19 Sept 1992, p.33-37.

(3) RANDALL, A. J. Resistance to new technology in the English industrial revolution. Resistance to new technology conference, Science Museum London, 1993. (To be published).

(4) RUCHT, D. The impacts of anti-nuclear power movement, an international comparison. Resistance to new technology conference. Science Museum, 1993. (To be published).

(5) HEALTH AND SAFETY EXECUTIVE NEWS RELEASE. More details of planned HSE study of health risks to women in semiconductor manufacturing industry. 27th May 1993.

(6) HEALTH AND SAFETY EXECUTIVE. The costs of accidents at work. HMSO 1993.

(7) HEALTH AND SAFETY EXECUTIVE. Successful health and safety management. HMSO 1991.

Safety and new technology in aviation – providing the regulatory framework

D A WHITTLE, BSc
Civil Aviation Authority, Gatwick Airport South, UK

SYNOPSIS Public transport aircraft accident rates have improved dramatically over the last few decades. There are many factors which have contributed to this improved safety record but there is no doubt that new technologies have played a major part.

In the United Kingdom, the Civil Aviation Authority is responsible for establishing the regulatory framework through which the industry can achieve high safety standards. This paper describes how the development of automatic landing systems provided the initial impetus for establishing a framework of objective safety requirements to support the certification of all types of aircraft systems, and how these have been adapted to take account of the widespread application of digital systems.

Developments in integrated systems are bringing new challenges to safety assessment practices and will need to be carefully evaluated by the industry and aviation authorities to ensure that the trend in improving safety standards is maintained.

1 INTRODUCTION

In all spheres of life, new technology expands people's horizons but sometimes this brings with it some additional risk to their safety. In aviation, the need to balance benefit and risk began with man's first attempts to fly. Today, air travel has become an indispensable part of our way of life and is generally accepted as being a very safe form of transport. This is confirmed by Department of Transport statistics (Table 1). This achievement owes a great deal to the continuing development and application of technology which has played a major rôle in contributing to improvements in the safety and performance of modern aircraft. As illustrated in Figure 1, the fatal accident rate of world scheduled services has been reducing steadily over the past 40 years. One of the major contributors to this trend has been the widespread introduction of jet turbine engines which brought a much needed improvement in engine reliability. Many interesting developments have taken place in the field of aircraft systems, such as automatic landing and 'fly-by-wire', and it is this area on which I wish to concentrate in this paper.

FATALITIES PER BILLION PASSENGER-KILOMETRES: Great Britain : 1980 - 1990	
Air	0.2
Bus and Coach	0.5
Rail	1.0
Car and Van	4.8
River and Sea	10
Bicycle	50
Pedestrian	72
Motorcycle	106

TABLE 1

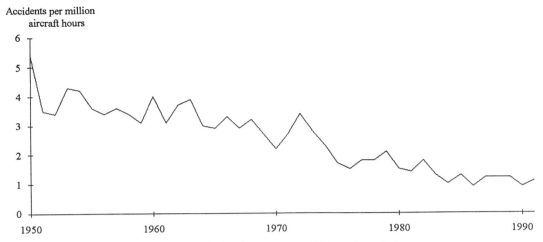

FIGURE 1 : Aircraft Accidents per million aircraft hours
Source : ICAO

2 RESPONSIBILITIES OF THE SAFETY REGULATOR

The United Kingdom Civil Aviation Authority has a statutory responsibility under the 1982 Civil Aviation Act to 'further the reasonable interests of the users of air transport and cargo services'. Safety is inherent in these objectives and is the most important element in them. Overall management of aviation safety is the responsibility of the CAA's Safety Regulation Group whose prime objective is 'to ensure that UK air safety standards are maintained and where possible improved'. This objective is pursued through a range of activities covering aircraft design and manufacturing standards, maintenance and operations, flight crew and engineer licensing, aerodrome licensing and air traffic services standards.

One of the principle tools employed to manage aircraft safety is the publication of airworthiness design requirements. In the past, these documents were known in the UK as British Civil Airworthiness Requirements, but through the European Joint Aviation Authorities, of whom the CAA is a founder member, these are gradually being replaced by Joint Aviation Requirements. It will be recognised by all engineers that to be valid and applicable to many different aircraft types, these design requirements need to be carefully written to achieve the safety objectives, but without unduly constraining the ingenuity of the designer. This is one of the driving forces which leads to close co-operation between industry and Authorities, something which in my view is essential to the successful implementation of practical and effective safety requirements.

3 SAFETY ASSESSMENT OF SYSTEMS

In the early days of the development of airworthiness requirements, new aircraft designs were often dealt with pragmatically, and many empirically derived requirements were written. However, as technology advanced, and systems became more complex, it became clear that a more objective method of evaluating the safety of a new design was required.

In the early 1960's, the UK industry set out to develop automatic landing systems. Here was a totally new concept for landing an aircraft without the normal skills of the pilot. The automatic landing system presented a major challenge to the certificating authorities. To have written a set of empirical requirements, such as the required number of processing channels, would have been too constraining on future developments. To deal with this problem, a joint industry/authority committee was set up, in the classic manner. What emerged was a means of stating the acceptable level of risk in probability terms.

The following quotation from the requirements which were ultimately published by the Air Registration Board (a forerunner component of the CAA's Safety Regulation Group) illustrates very well the change in philosophy brought about by the concept of automatic landing.

> *"It has not been usual to have to state a required level of safety in airworthiness requirements as they have generally been empirical means to improve in the future some feature found wanting in the past. In this case there is no past, so it is not a question of improving something which has been found wanting but of trying to ensure that a new feature will be acceptable. For designers to do this it is desirable to know what will be considered acceptable. Hence the need for a statement of the level of safety required."*

The 'level of safety' that was deemed appropriate for automatic landing was one which would 'not increase the risk of an accident'. This does not seem such a startling objective, but it began the principle of using historical safety performance as a basis for setting targets for the future. Significantly, it was also recognised that the safety target must not be set so high that the introduction of systems which would make a worthwhile improvement in safety would be prevented or unduly delayed.

The rate of fatal accidents in the landing phase was judged at the time to be around 1×10^{-6} per landing; that is 1 accident in 1 million landings. It was therefore decided that a target level of one order better, that is 1×10^{-7} per landing, should be set for the automatic landing system. Certification of the system, in addition to extensive performance testing, required the manufacturer to conduct a detailed failure analysis to show that this safety objective would be achieved. The first autoland (actually automatic flare only) in passenger service, with the Hawker Siddley Trident, took place in 1965 - a world's first.

4 CURRENT SYSTEM SAFETY REQUIREMENTS

The design of the Concorde aircraft brought about a need to extend formal safety assessment techniques to other complex systems such as the digital engine air intake control system. These generic requirements, which were subsequently introduced in British Civil Aviation Requirements, are now enshrined in the airworthiness codes of both the European Joint Aviation Authorities and the US Federal Aviation Administration (References 1 and 2). The underlying principle of these requirements is that the probability of failure and the resulting effect on the aircraft should have an inverse relationship (Figure 2).

The numerical basis for this requirement was determined by setting an objective of 1×10^{-7} per hour for a catastrophic accident due to systems failures, which was about 10 times less than the systems related accident rate at the time. This is a difficult objective for a manufacturer to deal with during design and therefore safety objectives are also defined for any one individual failure condition which would cause an accident. This is set at two orders less, that is 1×10^{-9} per hour, on the arbitrary assumption that there may be 100 such failure conditions in the aircraft. Examples of these failure conditions might be loss of pitch control or loss of flight instruments.

The intent of this requirement is that such a failure should 'virtually never' occur. To put it into terms which are more meaningful, in a fleet of 100 aircraft of a type, each flying 3000 hours per year, one of the catastrophic events might be expected to turn up once in 30 odd years.

It would be wrong however, to give the impression that safety assessment is all to do with probabilities. Compliance with sound engineering requirements which have been developed from many years of experience of incidents and accidents is a major factor in building safe systems. Safety assessment must therefore also address aspects such as single failures, independence of multiple systems and the potential for multiple failures from single events. Lightning strikes and engine high speed rotor bursts are two examples of single events which have the potential to cause a lot of damage to an aircraft if proper precautions are not taken.

Occasionally, as was unfortunately demonstrated by an aircraft over Sioux City in 1989, such precautions are found to be wanting. In this case, an engine burst in a manner which released far more energy than would normally be the case and the resulting systems damage left the aircraft with severely limited control. What was effectively a totally catastrophic failure, was only mitigated by the exceptional skill of the flight crew who managed to coax the aircraft into a 'controlled' crash landing. Such accidents are, fortunately, rare, but the industry acts upon them and where necessary takes steps to reduce the risk of similar events occurring.

5 DIGITAL SYSTEMS

A paper of this nature would be incomplete without some coverage of digital systems. This technology has provided

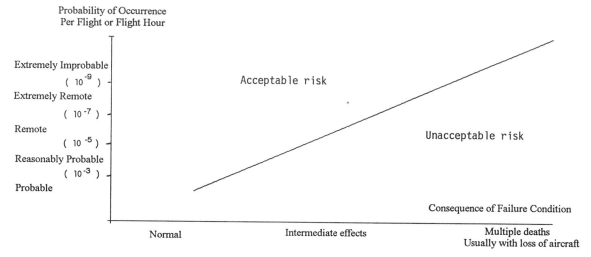

FIGURE 2 : Relationship between the consequence of failure and the probability of occurrence

the basis for an explosive growth in the capabilities of aircraft systems. (See Figure 3). In recent years, media headlines have focused on the 'glamour' systems of 'fly-by-wire'. It is important to bear in mind, however, that the underlying technology has been employed on aircraft in various applications over many years. Other examples include autoflight systems, flap controls and wing load alleviation as used on the Lockheed L1011 aircraft. This experience has shown that a step-by-step, evolutionary approach to such developments can be achieved safely. However, the single most important factor embracing this experience is the emphasis on the 'total system' concept. This is not a new concept but its significant is perhaps more pronounced where software is involved. The key is that the software is but one element, or more often several elements, in the total system design. It is important to always keep in mind the functions performed by the system as a whole and the safety requirements inherent in performing those functions. A good system engineering design organisation will then make appropriate use of the technology available, taking account of his experience and the characteristics of the available technology. As far as the software itself is concerned, a tremendous international effort has gone into the development of acceptable standards, the most recent version of which was published in late 1992 (Reference 3). This has certainly required a new approach, since no practical means has been developed to carry out a 'hardware type' failure analysis. The technique which has been adopted requires software to be developed in a systematic and rigorous manner with the amount of rigour to be used in the process being determined by the contribution of the software to the system failure effects at the aircraft level. In safety critical systems, therefore, it is important to minimise the potential effects of software design errors by selecting a suitable total system architecture.

6 SAFETY VERSUS PERFORMANCE

New technology can often be exploited for the benefit of safety or performance; sometimes both. Earlier in the paper, I mentioned the major contribution to safety of jet turbine engines. The combined reliability and performance of current engines has now reached the point where long over water flights are routinely conducted by twin engine transport aircraft.

If I was to cite one example of a system which has made a significant contribution to safety, it would have to be the Ground Proximity Warning System. This device, which is mandatory for larger aircraft in most western states, has undoubtedly saved many aircraft from being flown into the ground since its introduction in the 1970's.

One of the most recent developments which offers a complex mix of benefits and safety issues is fly-by-wire. The aircraft manufacturing and airline industries are intensely competitive and it is difficult to imagine a manufacturer investing in fly-by-wire technology unless there was seen to be an economic benefit for doing so. However, it is not my place to comment on that aspect, nor am I qualified so to do. Nevertheless, fly-by-wire does offer potential safety benefits such as stall avoidance, structural overload and flight envelope protection. To achieve these benefits, however, it has to be recognised

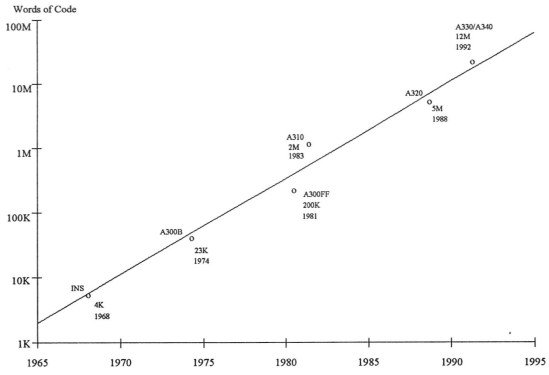

FIGURE 3 : Growth in Civil Airborne Software (Courtesy of Airbus Industrie)

that some additional effort, which may be considerable, has to be invested in satisfying the certification authorities that such a system is at least as safe as conventional controls.

7 FUTURE DEVELOPMENTS

The aircraft industry has moved rapidly in the last twenty years to take advantage of technological developments. Some of those associated with aircraft systems are shown in the chart in Figure 4.

The underlying technical theme of these developments is systems integration. Aircraft functions need no longer be performed by dedicated systems. Cockpit displays are a simple example of this trend. Modern aircraft no not have individual instruments for airspeed, altitude and attitude. Instead these are presented on a common electronic display, with back up displays in the event of failure.

This increasing level of integration will inevitably continue, driven primarily by perceived performance and economic benefits. One effect of this trend will be an increase in the complexity of system safety assessment. The issues involved in the development and certification of highly integrated systems are currently being studied by an industry committee which has been tasked with developing guidelines on the subject.

The expansion of systems integration, however, undoubtedly holds the key to two promising developments which will aid both performance and safety. The first of these has become known by the generic acronym of ESAS (Enhanced Situational Awareness Systems). The main driving forces behind this concept are the continuing occurrences of 'controlled flight into terrain' (CFIT) accidents, and the desire of many airlines for radar or infra-red enhanced vision systems as a cheaper alternative to fully automatic landing systems for low visibility approach and landing. Current technology Ground Proximity Warning Systems do not have a look ahead capability and therefore are blind to steep mountainsides. This deficiency can be overcome by the use of a forward looking sensor, or by use of a detailed map database which can give a pre-warning of the proximity of high ground. The second development involves integration on a much larger scale. Management of air traffic today is totally dependent on voice communication which as everyone knows is a fallible medium. Major research and development programmes are now underway to provide an air traffic management system which will employ air/ground data links, directly or via satellite, to carry both the air traffic controller's instructions and the aircraft's position.

Both of these developments will present many new safety issues which will need to be resolved.

8 HUMAN FACTORS

I mentioned earlier that the proportion of aircraft accidents directly attributable to systems failures was round about 10%. By far the highest proportion is due to human failure, such as 'crew error'.

When considering the safety issues associated with new technology, it is essential to take account of the human element, whether it is the designer, operator or maintainer. If we are asking the human to do something different to what he has been used to doing, or trained to do, we are inviting a potential problem. Where radically new cockpit controls and displays are introduced, we need to ensure

↑ Increasing integration

Data link

Integrated Modular Avionics

Fly by wire

Integrated displays

Flight Management Systems

Active Controls

↑

Automatic landing

time now

Powered flying controls

Discrete Systems

time →

FIGURE 4 : Growth in Systems Integration

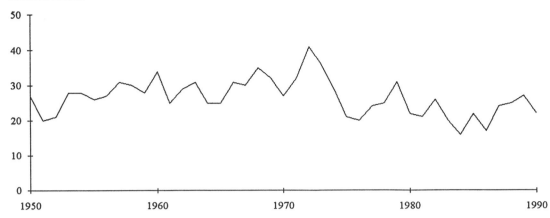

FIGURE 5 : Aircraft Accidents Scheduled Air Services, excluding USSR
Source : ICAO

that the human factors issues are thoroughly investigated. This is an area which requires much more research and understanding so that human performance can be properly included in the total safety assessment of new systems.

9 CONCLUSIONS

Earlier in the paper, I showed how aircraft accident rates had maintained a downward trend over the last few decades. This, of course, is very encouraging but there are many reasons not to be complacent. It is generally recognised that the perceived risks of flying have little to do with the accident rate as such, but have a great deal to do with the frequency of accidents. For example, two major accidents occurring within a short space of time can bring about tremendous pressure for action by 'the Authorities'. It is the declared objective of the CAA, to seek a continuing reduction in accident rates so that growth in air traffic does not increase the frequency of accidents. As Figure 5 shows, this has broadly been the case in the last forty years.

Technological advances will undoubtedly offer opportunities to continue improving safety but we must ensure that in harnessing those benefits, we do not unwittingly create new hazards.

It is a hallmark of the aviation industry that communications, co-operation and sharing of problems is recognised as a vital ingredient of the safety culture and this can only be achieved successfully when safety receives the commitment of management at the highest levels. This is probably the most important component of the framework through which effective safety regulation can function. I hope that this paper has helped to contribute to that process.

10 ACKNOWLEDGEMENT

I would like to acknowledge the support and assistance of the Civil Aviation Authority in the preparation of this paper. However, the views expressed are solely attributable to the author and are not necessarily those of the CAA.

11 REFERENCES

1. Joint Aviation Requirements
   JAR-25 Large Aeroplanes

2. Code of Federal Regulations
   Part 25   Airworthiness Standards : Transport Category Airplanes

3. Software Considerations in Airborne Systems and Equipment Certification
   EUROCAE ED12B/RTCA DO-178B

Management responsibility for the safety of software

D W NEWMAN, CEng, FIEE, MBCS
Ford Motor Company, Dunston, UK

SYNOPSIS A dramatic growth in the use of microprocessors and software to implement critical control systems is taking place within our society. There is now greater media and public awareness of the benefits and disadvantages associated with such systems used in our every day life. As car drivers, the general public will probably experience more direct contact with safety related software than through any other every–day application. It is, therefore, vital for the automotive industry not only to get it right but be seen to be getting it right.

This paper does not attempt to summarise these standards, rather it highlights some aspects that are of particular relevance to managers responsible for the development of these systems.

INTRODUCTION

As microprocessors are introduced into a broad spectrum of engineering applications, we see many electrical, electronic and mechanical engineers adapting and being re–trained to produce compact, responsive and cost effective solutions closely tailored to the applications familiar to them. This gives an opportunity to compare aspects of established electrical and mechanical reliability engineering to the recommendations offered in recent software standards for safety related application.

CONTRASTS WITH TRADITIONAL COMPONENTS

Compared with the traditional forms of engineering, software has some differences and similarities:

The Differences

- Software is primarily a design and has no manufacturing variations, wear, corrosion or ageing aspects.
- It has a huge capacity to contain sophistication and complexity. For example, in its ability to provide direct and indirect communication with the user, which is important when the hardware begins to deteriorate or fail.
- It is invisible.

The Similarities

- Software can be developed using accepted large scale engineering practice and can be continuously improved.
- It has quality attributes akin to other engineering activities.
- It can fail in the sense of being inadequate for its purpose and needs to be tested and validated.
- It requires a variety of specialist skills that need to be develope through training, and then managed effectively.

IDENTIFYING BEST PRACTICE

Micro–processor and software technologies are continually and rapidly evolving, compared with other more traditional engineering disciplines. However, some of the concepts are far from new. Structured Analysis and Modularity, for example, were first introduced and accepted as part of software engineering in the 1960s. Difficult judgements are involved in establishing what is todays *best practice* for developing future critical products. Fortunately, there are a number of sources of help available:

- The Institution of Electrical Engineers has published a *Professional Brief* on safety related systems (1)

- The British Computer Society has recently published a policy statement (2).
- The Department of Trade and Industry in association with other government departments is funding research under the title of *Safe–IT*. A consortium of the UK automotive industry is preparing guidelines in association with this programme.
- Much of this work is associated with the International Electro–technical Commission's draft generic standard (3).

Encouraging and enabling staff to participate in the development of standards and attend the related conferences is equally important and can be considered as advance training. DTI and SERC in conjunction with IEE and BCS have organised a Safety Critical Systems Club that has regular seminars throughout the UK and publishes a newsletter. Recent papers have been published in paperback. (4&6).

MANAGEMENT RESPONSIBILITIES

Management responsibilities can be identified that have strong influence on the principles behind many of these best–practice techniques, including:

- Understanding and specifying the requirements
- Defining the scope of *software* and *control systems*
- Organisation of software to handle complexity
- Resourcing for *diversity*
- Understanding any special factors related to the application

UNDERSTANDING & SPECIFYING THE REQUIREMENTS

One of the most difficult stages in the development life–cycle is understanding and agreeing the overall requirements and recording them in a clear and unabiguous manner. A variety of well established methods and techniques are in use, from plain narrative text, diagrammatic and tabular techniques though to mathematics. However, the written word remains the clearest medium for communication between product planners, engineers, managers and customers.

Whilst much critism has been aimed at English narrative for the introduction of ambiguities into the functional requirements specifications of computer systems, it remains the most effective medium for defining overall requirements such as objectives, design concepts, quality, risks, safety, etc. The English language, in particular, is gaining in worldwide status. Too often, however, examples of verbosity and ambiguity are used as justification for resorting to other methods that give lesser clarity to the wider community, or worse still, are used for justification of a void.

DEFINING THE SCOPE OF SOFTWARE AND CONTROL SYSTEMS

The perceived reliability of software in any application is influenced by factors and disciplines that go way beyond those generally accepted within the narrow confines of *Software Engineering*. This can be viewed as Software Engineering in the centre of a wheel surrounded and influenced by other disciplines and control systems technologies as shown in Figure 1. All can have quality, reliability and safety impact.

It is more important for Management to ensure that the appropriate culture, disciplines and resources are in place throughout systems development, rather than to attempt to establish a narrow boundary of *safety critical software* in some arbitrary manner.

THE ORGANISATION OF SOFTWARE TO HANDLE COMPLEXITY

The ability to contain complexity within software is limited only by the ability to manage the human communications and human resource aspects of that complexity. The problems and the solutions are not unique to Software Engineering. Openness by visualisation of structure, clarity of top–level design, with well defined interfaces provides one key to the solution. There are many diagrammatic techniques used in industry to convey *structure* of organisations, mechanical systems and processes. Suprisingly, few of these are used to visualise the structure of control systems and software. One problem is the lack of physical boundaries. This can lead to indecision on where to establish the formal, but sometimes arbitrary, interfaces.

RESOURCING FOR DIVERSITY

There is a good analogy between *hardware redundancy* and *software diversity*. In the case of hardware, the statistical methods for calculating the benefits of duplication, triplication, or more, using known component failure rates are well established. See figure–2.

If common–mode–failures can be reduced to insignificance, calculations can show that duplication gives not just an improvement of a factor of two in system reliability, but can give a dramatic improvement of the order of 100:1 or more!

Estimating the level of human resource to achieve adequate diversity, is, however, more difficult. Fortunately, the above analogy does indicate the scale of benefit that can be achieved. If similar statistical methods for random error and failures in software can be applied, then similar orders of magnitude of software reliability improvement could potentially be achieved for a modest level of software diversity. *N-version programming* does not offer all the answers. In some applications, there is now documented evidence (5) collected over many years of operational use that this technique does not give the returns on reliability that were once predicted. There are too many analogies between the *common mode failures* which frequently reduces the hardware reliability achievements and the *independance* levels associated with software. The use of software validation tools, the organisation and separation of staff and deployment of independant technical audits, are management/planning judgements with direct impact on the effectiveness of diversity and hence software quality, reliability and safety.

SPECIAL FACTORS IN AUTOMOTIVE ENGINEERING

Some special factors separate software in automobiles from the nearest similar application in, for example, civil aviation. These factors relate to the volumes and component spread that vehicle software has to handle in comparison to aircraft, where that industry has full control over conditions of use and maintenance standards.

As the functions controlled by software broaden in scope and take on greater criticality, the importance of handling hardware component failure and deterioration also increases. The power and capability of microprocessors and software enables sophisticated fallback and fail-safe techniques to be viable on even the smallest systems. However, difficult judgements on what is safest, in apparently *impossible* situations, must be made at the design stage. This is particulary true in situations where there is no trained pilot, operator or company employee available to take over manual operations. To abdicate and not provide the mechanisms is often perceived as system and software UN-reliability by customers, rather than the root-cause hardware failure or hardware deterioration.

IS ABSTINENCE THE ANSWER?

It is not unknown for these issues to be avoided by a strategic business decision not to use software in a safety related role. Perhaps, this is not surprising when it is recognised that software cannot be 100% validated as *error free* and that the designers could be liable for any omissions identified over the whole life of the product. Equally however, the overwhelming benefits in reliability and consistency, that are being achieved in comparison with the alternatives, must now be recognised. General acceptance of these achievements must soon result in recognition that to NOT utilise these benefits in saftey related applications, will be seen as the greatest omission of all and a failure to adopt best practice.

CONCLUSIONS

Many of the techniques and lessons learnt with electro-mechanical systems in demanding high-reliability environments have relevance to the development of Software. Consequently, many of the traditional engineering management responsibilities such as basic design assumptions, the encouraging open communication, organisation of resources and training are equally relevant and also important. In safety related applications, however, the complexity issues and relative invisibility of software make these responsibilites even more important.

REFERENCES

1. *Safety-related Systems*, IEE Professional Brief, January 1992
2. *Policy Statement on Safety Critical Computer Systems,* The British Computer Society, 1993
3. *Software For Computers in the Application of Industrial Safety-related Systems.* IEC SC/65A/122, September 1991
4. F.Redmill & T.Anderson. *Safety Critical Systems Current Issues, Techniques & Standards Ed.*: Chapman Hall Jan. 1993
5. F.Redmill & T.Anderson. *The Swedish State Railway Experience with N-version Programming. Page-36 of: Directions in Safety-critical Sytems.* The Proceeding of the 1993 Safety Critical Systems Symposium. Ed.: Springer-Verlag 1993

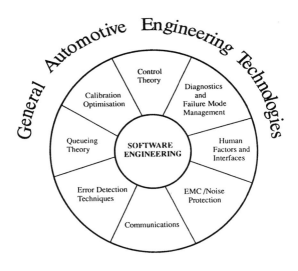

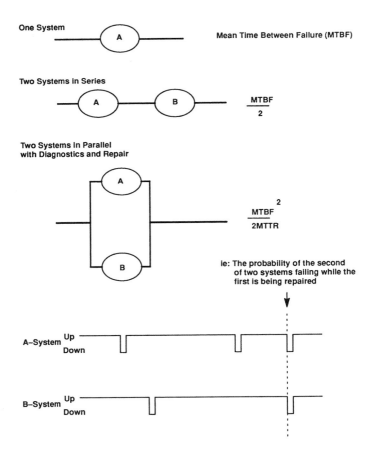

Safety management of the Sizewell B project

B V GEORGE, FEng, FIMechE
Nuclear Electric plc, Knutsford, UK

Introduction

The Consents and Licensing Processes

I will start by outlining these processes for Sizewell B since they will be unfamiliar to most of you. I cannot think of any plant, nuclear or otherwise, that has had the level of searching exposure and examination of its safety case as Sizewell B. The design has been subjected to 2 major public inquiries, together with a high level of scrutiny by HM Nuclear Installations Inspectorate (NII) of the Health & Safety Executive. The Sizewell B Inquiry opened in January 1983 and did not close until March 1985. During that period there were 340 days of hearing, over 16 million words of evidence and a total of 195 witnesses.

The consents & licensing processes are illustrated on Figure 1, including some of the inputs, up to the granting of consent by the Government on 12 March 1987 and the issuing of the Site Licence by the NII on 4 June 1987.

Following the consents process and the issue of the Site Licence, we have proceeded through a number of key safety stages (Figure 2) at which we have had to demonstrate to the NII that we were vigorously implementing all of the safety commitments given in the Pre-Construction Safety Report. We have successfully passed all of the licence construction & commissioning stages to date. We also provided the NII with a full Pre-Operational Safety Report consisting of 41 volumes, including source documentation, in November 1992, 12 months ahead of fuel load, to give the Inspectorate plenty of time to assess the total safety case before we ask for consent to load fuel.

The licensing process has been extraordinarily onerous. On the other hand, I believe that the processes as applied to Sizewell B have been better managed and can be demonstrated to be more comprehensive than any previous licensing process in this country. Without this rigorous monitoring & progressing of safety matters by these intermediate safety stages, I would not be confident that we would be given consent, in a timely manner, to load the fuel into the reactor.

We invited a team of international experts assembled by the International Atomic Energy Agency to carry out a Pre-Operational Safety Review (Pre-OSART) of the commissioning and early operational phase of Sizewell. That was a most successful exercise, provided valuable input and tended to confirm our view that we were, if anything, ahead of the field in safety terms.

The Use of Computers

Let me turn now to the use of computers in safety-related and safety-critical systems at Sizewell B, and to discuss some of the unique project management problems that we have had to tackle as a result of this.

Computers are being used increasingly for safety-critical applications in major projects. Managing the production of software for these systems is fraught with potential difficulties. Software engineering is a young and still rapidly changing field with few established customs and practices. Methods of estimating how long, and how much effort, will be required to produce the software for a system are not well developed, and there is little project management experience to draw on. In addition, standards against which the quality of the final software is judged are evolving rapidly, and can change significantly during the timescale of a major project.

All of these difficulties have been tackled - and overcome - for Sizewell B. I will try to explain how, and the lessons we have learned in the process.

The Control & Instrumentation (C&I) Systems

The C&I systems of a nuclear power station are usually thought of in 3 broad functional groups:

1. the automatic control systems which regulate the operation of the plant under normal circumstances,

2. the manual controls and instruments in the control rooms which allow the operating staff to supervise operation of the station, and to take corrective action if required,

3. the reactor protection system which trips the reactor and starts safety

features if a fault occurs.

On Sizewell B, each of these functions uses computers to some degree. The rest of this paper will be about the Station's Primary (reactor) Protection System (PPS), which employs state-of-the-art digital technology. In addition to the computer-based PPS, the Secondary Protection System and a set of hard-wired controls and displays provide diverse safety back-up that is based on tried and trusted technology.

Reasons for using Computers in Safety-critical Applications

We chose to use computers in safety-critical applications because they offer significant benefits:

i. The quantity of small signal cabling is reduced by the use of multiplexing datalinks. Thus less cables have to be installed, and this also reduces the quantity of combustible material present on the Station.

ii. Configuration of the systems for their specific tasks is done by programming rather than site wiring. This means that the cubicles can be installed unconfigured, and then have their software installed later, and that the process of writing the software can take place in a better environment than on the job site.

iii. The performance of computers, in terms of their speed of response, accuracy, freedom from drift, and ability to process algorithms, can be much better than conventional analogue systems.

iv. The interface between the operators or maintainers of the systems can be made more user-friendly, improving the quality of operation & maintenance and reducing human errors.

v. Computers can detect corruption of signals due to fire or interference, and can also test and calibrate themselves while in operation, alerting operations & maintenance staff to failures before these cause an operational problem.

Project Management

Turning now to the principal topic of my presentation, managing the timely production and delivery of the computer-based Primary Protection System (PPS), I will describe the steps that we have taken to ensure that the software is of the required high quality, and is delivered to the site on time.

The software design and implementation process for the System evolved in the early 1980s, with the object of producing code of the highest quality which is easily maintainable, verifiable and re-usable in a number of applications.

Figure 3 shows the development programme of the PPS, indicating the separate but mutually dependent activities of design, manufacture, testing and licensing. In addition, towards the bottom of the figure, are shown the unique activities of dynamic testing and the independent engineering and MALPAS assessments that we have performed to ensure the safety of the System. I will describe these activities in a little more detail so that the difficulty of including these within the programme of a major project can be appreciated.

The manufacturer of the PPS, Westinghouse, performed an extensive programme of software verification to detect and remove errors generated in the design and coding phases of the project. This included the review of all the documentation, source and object code for the System by an independent in-house review team. Notwithstanding this extensive quality assurance effort by the contractor, we put in place our own, very extensive, independent software verification & validation activities. These activities are summarised on the figure and, briefly, comprise the following:

Fitness for Purpose Review

1. Engineering Confirmatory Analysis

    This includes the formal review and approval of top level documentation such as the System Design Requirements, Software Design and Coding Standards, Software Design Requirements, by Nuclear Electric, and a line by line review of the Source Code by NNC Limited.

2. Independent Design Assessment

    The principal design documentation was assessed by Nuclear Electric's own Independent Design Assessment Team. This Team was involved at an early stage in the design process, prior to final verification, and therefore findings from the assessment process could be incorporated into the design.

Confirmatory Supporting Activities

1. MALPAS Analysis

    The source code is analysed using the Malvern Program Analysis Suite (MALPAS) of static analysis tools. An important aspect of this tool is its ability to develop mathematical relationships between inputs and outputs and compare these automatically with the specification. The MALPAS analysis of the source code enables the assessor to "close the loop" between the specification and the code.

2. Source/Code Comparator

    In order to "close the loop" between the source code

and the PROMS containing the delivered software, a comparison tool has been developed. The data in the PROM programming tapes is translated back into assembler language which is then compared with the source code using the MALPAS Compliance Analyser.

Testing

We have also embarked on an extensive programme of testing to provide empirical evidence that the PPS will perform as specified. A summary of the key testing activities is shown on Figure 4.

i. Testing during the Design Phase

   Design testing was an explicit phase in the development of the PPS software. Sophisticated test tools such as in-circuit emulators and the full-scale prototype were used in this process.

ii. Testing during Manufacture

   Following hardware and software integration, each of the four sets of PPS equipment was subject to extensive Factory Acceptance Tests to prove compliance with the specified functional and performance requirements, and a 400 hour soak test was carried out on the equipment.

iii. Site Testing

   Following the installation of the equipment at site, a system test is carried out similar to the Factory Acceptance Tests, and an integration test is performed to demonstrate the performance of the completed system. In addition, for an extended period of up to one year prior to fuel load, the complete PPS will be operated under representative conditions.

iv. Dynamic Testing

   To give additional confidence in the performance of the PPS, a programme of dynamic testing was performed. Specific functions of the PPS were targeted, including those functions where the highest reliability is claimed. Automation of the testing allowed a large number of test cases to be simulated, and over 50,000 tests were performed.

Management of the Process

Interfaces between different suppliers can be a nightmare, because of the difficulty of getting two different computer systems to talk to each other, especially when they are built to proprietary standards, and commercial interests are a risk. It is all too easy for a dispute to become bogged down in technical detail. This is a lesson we learned from previous projects. For Sizewell B, we have deliberately divided the work into major packages, and have given each of these to a single, experienced, established, contractor who is also the designer and manufacturer of the equipment.

Our involvement in the process has been that of an informed customer, monitoring and guiding the contractor's progress. We believe that it is important to have a very clear and high quality specification and programme, and effective ways of monitoring progress against these. It is also vital that there is a full and frank exchange of information between the two parties. We have found that an "exchange of hostages", where we placed two Nuclear Electric representatives with the Contractor, and the Contractor placed two with us, is particularly valuable. These representatives work in the design teams, and form an open channel for communication, and facilitate the exchange of knowledge and cultures.

<u>Lessons learned</u>

1. The delivery of the software is just as critical as that of the hardware. Although this may seem self evident, the intangible nature of software, and of its development process, means that it is much easier to measure and monitor production of the supporting hardware, and that the software can be easily overlooked. The complexity of software is much greater than that of hardware.

2. The current absence of well developed and stable procedures for producing software to safety-critical standards means that the flexibility and amenability to change of any safety-critical system employing software is lower than for hardware-based systems. This has several knock-on effects:

    i. The design details have to be frozen earlier than would be expected because the development cycle is long, due to the complexity of software, and because of the time required for verification & validation. The result is that changes cannot be accepted late in the day without repeating all of the preceding verification & validation exercise, knocking the programme off course.

    ii. The apparent ease with which programmable systems can be changed or reconfigured is very tempting to system designers, as is the complexity of the algorithms that can be implemented. This has to be resisted. The functionality of the systems must be kept as simple as possible to minimise the development problem, and also the verification & validation problem. The history of software development is littered with projects that have not come to fruition because of over-ambitious goals.

3. It is important to select a single, large, competent and experienced contractor as the principal supplier. Such a contractor has a detailed knowledge of the systems to be interfaced, and can anticipate and overcome difficulties that would frustrate third-party suppliers.

4. Because of the intangible nature of the software product, it is important that a clear and detailed specification is established up front that is agreed by both parties. An "exchange of hostages" can help to ensure a free flow of information between the two organisations. We have found it important to conduct regular in-depth design reviews, using our own technical experts and independent consultants, so that we can be sure that the progress being reported is being achieved. It is also vital that the project management organisation does not become bogged down in the details of the production process. It is important to be aware of these details, but also that we can stand back and see the big picture, and then use "carrots & sticks" to guide the overall progress of the contractor.

5. Managing the production of computer systems, particularly safety-related ones, is different from managing conventional hardware projects. Traditional project management techniques often do not work when dealing with software development. This is due to the complexity of a large software project, which has many interacting and mutually dependent components. It runs directly counter to the traditional "pour in more resources" school of project management, and illustrates why it is important to have people with relevant experience on the team.

Concluding remarks

It is possible to manage a major project successfully that includes computer-based systems as critical components. However, the absence of experience, standards and stability in software development means that this area can be pivotal to the success of the project, and can have an effect that is grossly out of proportion to its capitalization in the venture. The very real benefits in safety and availability that can follow from the adoption of computer-based systems mean that, in the future, most successful projects will employ computers in one form or another. If we do not learn how to tack the problems successfully, our competitors will.

Sizewell B is one of the most successful nuclear power station projects undertaken. It is well ahead of programme and under budget with less than a year to go to commercial operation. We are proud of the fact that we have achieved this with a reactor system new to the UK, and using some of the most advanced computer technology on any power station in the world.

I hope that some of the techniques that we have used, and the lessons that we have learned, will be of help in your work.

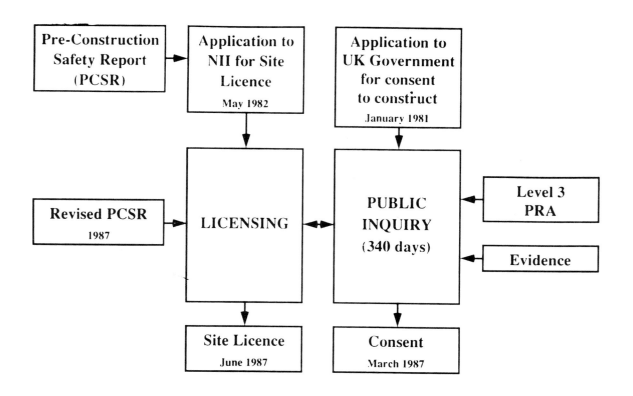

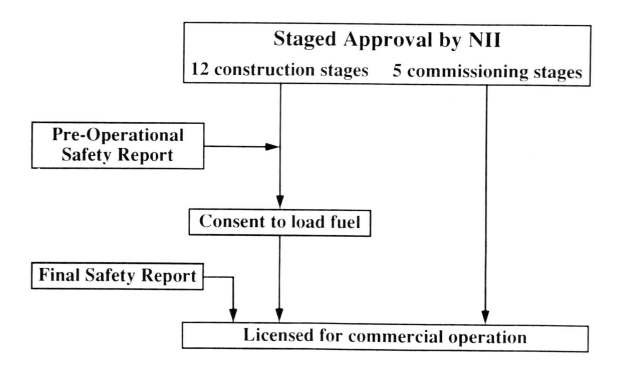

Fig. 1 Scope of vehicle based software

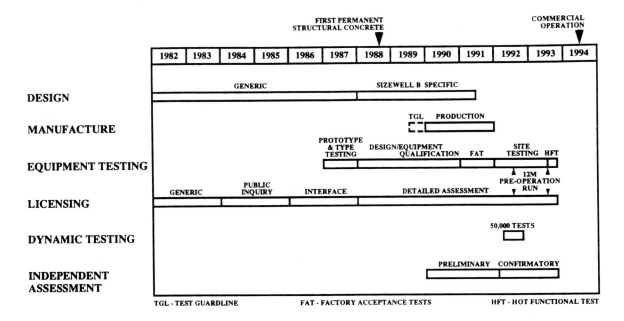

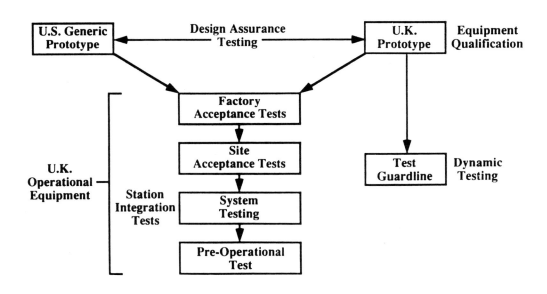

Fig. 2 Component failure rates

Legal consequences of accidents – managing the product liability issue

A J HOBKINSON
Cameron Markby Hewitt, London, UK

INTRODUCTION

The concept of management for safety is a relatively new one for the legal profession who have until recent years traditionally been viewed as a necessary and expensive evil to be referred to as a last resort. While lawyers are still regarded as expensive, the necessity to involve them in product integrity issues has gradually been realised. Product liability lawyers are no longer confined to claims handling but increasingly are becoming involved in loss prevention and disaster planning.

This paper illustrates some of the legal exposures manufacturers and suppliers face and the potential consequences of their errors. In doing so it will help demonstrate why preventative medicine is becoming increasingly popular.

PRODUCT LIABILITY

The liability of product manufacturers, and in many cases therefore also the sales intermediary, will usually fall into one of the following three categories:-

1. Defective manufacture: a failure of the manufacturer to meet its own design specification.

2. Defective design: where the product has been manufactured in accordance with its design but the design itself is flawed.

3. Inadequate warnings or instructions for use: this is particularly relevant to products that are inherently hazardous, such as drugs, a number of D.I.Y. products, and electrical appliances.

These categories are not necessarily mutually exclusive, but where both defective manufacture and design are in question, it will usually be easier for a consumer to prove the former than the latter. Indeed, disputes over design and the adequacy of warnings are ultimately matters of expert evidence and are therefore fertile ground for protracted and expensive litigation.

A breach by manufacturers in any of the above areas may give rise to claims or costs in one or more of three distinct but related areas:-

(a) Damage or injury to third parties or their property.

(b) Damage to the product itself.

(c) Product recall.

The distinction between these categories is important in two fundamental respects. First, the nature of the damage will determine the legal exposure of the manufacturer. For example, the concept of strict liability introduced by the 1985 EC Directive on Consumer Safety[1] does not extend to damage to the product itself. Secondly, while product liability insurance policies will indemnify a producer in respect of third party claims, they will not usually reimburse the manufacturer for the cost of repair or replacement of damaged products. Further, while product recall insurance is available, it is expensive and usually subject to strict financial limits.

THE LAW

There are three primary categories of legal liability under English law, namely breach of contract, tort (negligence) and statutory liability. In practice, all three are often present in product liability claims.

1. **Breach of Contract**

    Most product liability claims will involve the examination of a contract. The contract need not be in writing, although it usually is, often on the standard terms of either the supplier or the purchaser. In specialist industries such as aviation, products are supplied under a specifically negotiated contract reached after discussion between the two parties. This type of contract will normally contain a detailed technical specification.

    In most contractual claims, there will be more than two parties involved because often there is a chain of supply between the manufacturer and the end user. Staying with the aviation industry, a component manufacture such as Lucas, for example, may supply parts to British Aerospace or Airbus. These parts will be incorporated into a larger product which is then sent on to the customer, such as an airline. It is also not uncommon for component part contracts to include an agreement that the benefit of any warranties will be passed on to the ultimate purchaser. The contract between Airbus and the airline, therefore, may be as relevant to Lucas' exposure as the contract between Lucas and Airbus.

    The following is a short summary of certain salient points in contractual disputes:-

    (i) A supplier of goods to a consumer is strictly liable to that consumer for any injuries caused by a product which is defective, irrespective of whether that supplier could himself have discovered that defect.

    (ii) The supplier's liability to the consumer for death or personal injuries cannot be excluded or limited.

    (iii) The consequential liability of the manufacturer to indemnify the supplier may be modified or excluded by contract, but where they are dealing on one party's standard terms any exclusions or limitations are only effective if they are reasonable.

2. **Negligence**

Unlike claims in contract where there is strict liability, in claims in negligence there must be an element of fault for damages to be awarded. In summary, the claimant must establish there was:-

(a)   A duty of care owed to him by the defendant;

(b)   A breach of that duty;

(c)   That the breach gave rise to the claimant's loss.

In general terms, a manufacturer will always owe a duty of care to the general public to ensure that his product does not cause physical harm due to any defect[2]. In the case of an aircraft accident, therefore, passengers will often sue not only the airline, but also the aircraft manufacturer, and the manufacturer of those component parts that are suspected to have been responsible for the accident. Indeed, there is often good reason for the passengers to sue the product manufacturers in that they, unlike the airlines, will not have the advantage of the financial limits of liability that are contained in airline tickets. These usually limit an airline's liability to 100,000 special drawing rights, which is presently equivalent to approximately £85,000. As the claimants are not parties to the contract of sale by the manufacturer, any financial limitations in the contract between the component supplier and the purchaser cannot be used in defence of the passenger claims.

The only defence the component manufacturer will have, therefore, is to show that he was not negligent, or in other words that he acted with all reasonable skill and care in the design, manufacture and supply of the product.

In most product liability claims there will be no direct contract between the end user or the injured party and the manufacturer. In those circumstances if it can be shown that the manufacturer failed to exercise reasonable skill and care in the manufacture or design of the product, the end user nevertheless has the option to sue the manufacturer directly in negligence.

Although in the case of businesses it is perhaps still more common for the purchaser to sue the supplier with whom it has a contract, in those circumstances it is also quite usual for that liability to be passed back up the chain of supply to the manufacturer himself. Even if a manufacturer is not negligent and therefore does not have a direct liability to the end user, this is a factor which must be borne in mind at an early stage of any product liability claim. It may ultimately prove less expensive to settle the claim up front than wait for it to be passed up the contractual chain.

3. **Breach of Statutory Duty**

The third main area of liability is imposed by statute. One product-related statute that operates independently of contract and negligence is the Consumer Protection Act 1987. This imposes a strict liability on manufacturers, and in some circumstances suppliers of goods, for damages a product causes to private individuals or their property.

The two other statutes which probably have the biggest bearing upon any product liability claim in contract, however, are the Sale of Goods Act 1979 and the Unfair Contract Terms Act 1977. In broad terms, the Sale of Goods Act specifies that goods sold under certain contracts must be of 'merchantable quality'. The Unfair Contract Terms Act limits the extent to which a seller can vary or exclude these implied terms in its conditions of sale.

EC LEGISLATION

The past ten years have seen a number of new EC Directives which have or will have a direct effect on the potential liabilities of manufacturers and suppliers.

On 1st March 1988 the Consumer Protection Act 1987 gave effect to the EC Consumer Safety Directive[1] in England and Wales. This introduced strict liability of producers to consumers for injury caused by unsafe products.

In November 1990 a draft Directive on services liability[3] was introduced. Although the final format and implementation date have yet to be agreed, the Directive proposes to move the burden of proof in claims so that it is for the supplier of a service to disprove an assumed culpability on his part.

On 29th June 1992 a Directive on Product Safety[4] was adopted and must be implemented by Member States by 29th June 1994. The Directive places a general obligation on all producers not to place unsafe products on the market.

Finally, on 5th April 1993 a Directive on Unfair Contract Terms[5] was adopted and has to be implemented by 31st December 1994. Although this is similar to our existing 1977 Unfair Contract Terms Act, the Directive is wider in that it applies to all types of contractual terms, not merely limitation or exclusion clauses.

All the above are additional to remedies available to individuals under existing national legislation.

THE CLAIM

(a) Control

While it is of course essential for a product liability lawyer to be fully familiar with these and other relevant laws, it is in dealing with the immediate and practical consequences of a major disaster or product liability problems where the product liability lawyer must excel.

In any major loss situation at least the following parties will be involved:-

(a)   Technical staff.

(b)   Lawyers.

(c)   Insurers.

(d)   Press Relations Officer.

(e)   Main board.

From the defence lawyer's perspective, the control and co-ordination of these component parts at the outset is critical. The board will often be besieged by the press and television. In cases where component parts are concerned, the technical staff on the ground will be involved in face to face discussions with their opposite numbers in their commercial customer. At the same time, there is a natural tendency for the company to indulge in an internal post mortem with consequential memoranda flying between various departments. The result is a defence lawyer's nightmare and a plaintiff lawyer's paradise as all the resulting documents must be disclosed if litigation results, and members of staff can later be asked about any conversations if they have to give evidence at trial. An experienced plaintiffs' lawyer will of course be fully familiar with the above scenario and therefore concentrate his enquiries in the appropriate directions. It is thus no use hoping that certain key documents will be overlooked, and in any event a defence lawyer has a separate legal obligation as an officer of the Supreme Court of England and Wales to ensure that all relevant documents are disclosed even if they are prejudicial to his client's case.

The key to controlling such problems (they will never be eliminated completely) is to have a designated team of key individuals and a plan of campaign drawn up in advance. With clear chains of command and channels of communication, the chances of unnecessary own goals being conceded is thus minimised.

(b) **Key Elements of Planning**

Such plans should be tailored specifically to each company's needs and specialisations. There are, however, a number of basic principles which are usually relevant to all industries. These can be divided into two parts. First, general policies which should, it is hoped, be general business sense and therefore be implemented in most organisations already. The second category is perhaps less obvious unless a company has already had the experience of being involved in a major claim. The following are merely examples. The list is by no means comprehensive:

General

- Record key decisions, especially on safety matters

- Keep accurate written records of other key matters, such as customer's instructions, specifications, safety tests and so on.

- Maintain a central record of all such documents: avoid the multiplicity of a large number of "personal" files.

- Do not examine claims/problems in isolation: are there wider safety implications for the future?

Post Accident

- Avoid having too many "spokesmen" at too many different levels: designate one or two individuals.

- Do not indulge in an internal "post mortem" by memoranda.

- Do not speculate in seeking to establish the facts.

- Do not be pressurised into providing explanations until more relevant investigations are complete.

Above all, whatever plan is implemented, make sure that <u>all</u> personnel at <u>all</u> levels are aware of the company's philosophy. Internal seminars in which these policies and procedures are discussed should be held at least once a year and ideally every six months. It is usually sensible to employ the services of an outside speaker to participate in those conferences, so that the company has the benefit of an external perspective on its practices and how they compare with competitors.

Practical Considerations

In any claim, whether it is a mass disaster or not, a number of practical considerations will come into play at an early stage:-

(i) The manufacturer may want a quick settlement to avoid damaging publicity.

(ii) In considering settlement, the costs of defending claims will (or at least should) be weighed. There are not only the obvious costs of instructing lawyers but also the considerable internal cost of lost management time and productivity.

(iii) The company's insurers will have a major influence on the course of action to be adopted as they will usually foot the bill for the payment of claims and legal defence costs. It pays for a company to have a good relationship and understanding with its insurers in advance as this will speed up the decision-making process immensely. It will also be a great advantage in situations where the interests of the company and insurers may not be identical. A company convinced of its lack of blame for an accident and wishing to defend its product may for example not wish to see an early settlement of a claim even if it makes sense to the insurer in pure financial terms because of the defence costs which would otherwise be involved.

Conclusion

Faced with rising legal costs and increasingly sophisticated claimants' lawyers, including one prominent figure who has suggested English lawyers are negligent if they fail to advise their clients in respect of the possibility of suing in the United States, it makes eminent sense for manufacturers to expend greater time and money in ensuring their products are made to the highest possible safety standards. It is therefore in the commercial sector's interests to be proactive in product integrity and safety issues, rather than simply reacting to EC directives emanating from Brussels.

As accidents will, by definition, never be eliminated entirely, it is also essential to have agreed procedures and key personnel in place ready to react at a moment's notice to a crisis. The ability of that team to balance out all the legal and practical considerations will determine the extent to which a major incident will have a long term adverse impact on a company's reputation and trading performance.

REFERENCES

1. Directive 85/374 of 25th July 1985.

2. Donoghue v. Stevenson [1932] AC 562 (HL).

3. Com (90)482 Final-SYN Brussels, 20th December 1990.

4. Directive 92/59/EEC of 29th June 1992.

5. Directive 93/13/EEC of 5th April 1993.

6. Dow Chemical Co. v. Castro Alfaro 786 S.W.2d674 (Tex 1990); cert denied, 498 U.S. 1024 (1991).

The management of safety and emergency planning

G D KENNEY, BSc, MSc, PhD, MASSE, MAIHA, MBONS, MIOD
Cremer and Warner Limited, London, UK

SYNOPSIS The investigations into the King's Cross Underground fire and Piper Alpha disaster concluded that failures within the management system meant that those in charge of the site at the time of the accident were not as prepared or competent to manage the situation as they should have been. In many organisations emergency plans are viewed as the "last resort" in the planning of a facility. A new approach to identifying the need for emergency plans is suggested using the "design" and "residual" accident concept of the Norwegian Petroleum Directorate. All such plans must be cognisant of the vital role that communications play in properly controlling an accident and the vulnerability of the chain to failure. Audits of the emergency capability of an organisation must be carried out and focus on the entire process of identifying the need for an emergency plan on through to assuring corrective actions are implemented.

1 INTRODUCTION

As a result of his investigations into the Piper Alpha disaster Lord Cullen concluded significant flaws in the quality of the management of safety directly and indirectly affected the circumstances of various disaster events. Lord Cullen considered that as a result of:

- Senior management being too easily satisfied,
- (Management) adopting a superficial response, and
- (Management) failing to ensure emergency training was being provided,

that the "platform personnel and management were not as prepared for a major emergency as they should have been". In his recommendations, Lord Cullen noted that one of the objectives for preparing a Safety Case was to demonstrate that the Safety Management Systems of an operator were adequate to ensure the full and safe evacuation of personnel in the event of a major emergency.

This paper will use the Piper Alpha incident as a case study to illustrate what might be classed a traditional approach to emergency planning versus a suggested innovative approach that a typical safety management system of the 1990's might need to demonstrate in a safety case.

2 CASE STUDY

An outline chronology of the sequence of events that occurred on Piper Alpha which evolved from the evidence laid before Lord

Cullen is presented in Table 1.

Table 1 Outline Chronology of Events during Piper Alpha disaster (all times approximate)

Time	Event
21:45	Condensate pump B trips
21:50/52	Low gas alarm - Module C
21:52/55	Two CG turbines trip
21:55/59	Further low gas alarms - Module C Third CG turbine trips High gas alarm - Module C
22:00	Initial Explosion
22:04/19	Broadcast of Mayday and Abandonment 100+ gather in galley/reception FRC's begin rescue efforts (approx 20 survivors rescued) Other platforms continue to operate
22:20	Tartan riser fails and ignites
22:21/49	Tharos closes to within 60m and sprays Piper with fire monitors Tartan initiates line depressurisations Other platforms continue to operate FRC's continue rescue efforts (approx 20 more survivors rescued)
22:50	MCP01 riser fails and ignites
22:51/59	Survivors on helideck (174' elev) jump FRC underneath platform destroyed by fireball MCP01 receives instructions to blowdown Claymore receives instructions to shut-down FRC rescue efforts continue (approx 12-15 more survivors rescued)
23:00	Claymore riser fails and ignites
23:01/19	Platform begins to collapse FRC rescue efforts continue (final 7-10 survivors rescued)
23:20/30	Main Oil Line fails and ignites

Of the total of 226 persons on board the installation on the night of the disaster, 61 survived. In addition to the 165 fatalities that occurred amongst the POB, two rescuers died when the FRC they were manning was destroyed when the Claymore riser failed and ignited.

It was possible to ascertain the primary cause of death in 131 cases. The predominant cause was from the inhalation of smoke and gas which accounted for 109 deaths. In 11 cases the deceased died from drowning and in another 11 cases the cause of death was attributed to injuries, including burns.

3 TRADITIONAL APPROACH TO EMERGENCY PLANNING

In many situations emergency planning is viewed as the "last element" employed when managing the risks or consequences of a particular hazard. In other words we focus on all reasonably practicable measures to prevent the hazard from occurring in the first instance. After all reasonably practicable preventative techniques have been developed, we next focus on those measures which will control and/or mitigate the escalation of an incident. After having explored and implemented all such reasonably practicable control measures, we then turn to planning for the worst case scenario should the preventive and control mechanisms fail.

Table 2 Traditional Approach to Developing Emergency Plans

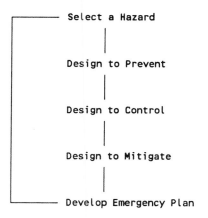

This philosophy is sometimes carried over into the allocation of resources to both install and maintain emergency equipment and/or emergency plans. In other words, it is not unusual to find that emergency equipment and plans are the last in line to receive funds in deference to other demands on scarce resources. There is a natural tendency where equipment, plans and procedures often sit for long periods without being called into action (and in some ideal cases may sit there for the entire life of a facility without being called upon) to consider them "necessary evils".

Even in organisations which have the best intents for being properly prepared to manage all reasonably foreseeable incidents that commitment will naturally decay over time. A number of factors influence the decay rate such as changes in personnel both in management and at the operational site, the absence of a company or site experiencing the "real thing" thereby contributing to a "can't happen here" syndrome. Hence the overall commitment and eventually the ability to respond to an actual emergency situation decreases leaving an organisation potentially exposed to criticisms similar to those levelled by Lord Cullen that management failures contributed to the platform personnel not being as prepared as they should have been.

3.1 Innovative approach to emergency planning

One concept which was well discussed in part 2 of Lord Cullen's investigations was the Norwegian philosophy of the "design" and "residual" accident event. One way to define a "design" accident is an event or incident where the design of the facility, plant, or piece of equipment is capable of breaking the escalation chain or sequence of events. A simple example of such a design accident is where a leak develops within a tank but the resulting pool is contained within a bund or drained to a safe location. A residual accident is one where the incident or event exceeds the capability of the facility to prevent further escalation without calling on "external" assistance. Returning to the leaking tank example a residual accident would be where multiple leaks occur in multiple tanks leading to overflowing of the bund or, possibly the leak from a single tank of such a size that the hydraulic head of fluid movement breaches the bund wall. These definitions for design and residual accident are similar to the concept of the "maximum credible event".

It is suggested that organisations begin their emergency planning process using the design and residual accident philosophy as opposed to the earlier described traditional approach. As depicted in the flowchart below, the process would start by defining a hazard then evaluating whether the facility can "absorb" the consequences. In many circumstances the ability to "absorb" an incident will require that certain hardware AND software measures be implemented. Hence I am not advocating that a "design" accident is dependent on engineering or hardware fixes or controls alone. Such software measures for an absorbed accident could be viewed either as a subset of the overall capability to contain an accident. Where the consequences of an accident or incident has the potential to exceed the design capabilities of the facility then it would be classed as a residual accident. As one would need some measure of the potential consequences to determine when an accident would be classed as residual, the magnitude of such consequences would act as the starting point for beginning to develop the required emergency plans. In a detailed examination of and development of emergency plans one might actually go through the loop twice, the second circuit would be focused on evaluating the "design" and "residual" accidents which a company's emergency plans could handle. One could couple this process into a quantified risk assessment of the potential for such mega-emergencies to determine the likelihood of occurrence and even possibly a cost-benefit analysis which would examine the costs to control such mega-events against the benefits.

Table 3 An Innovative or Predictive Approach to Developing Emergency Plans

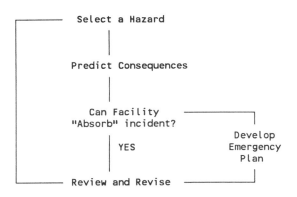

4 EMERGENCY MANAGEMENT PROCESS

The handling of an emergency situation can be viewed as a five step process:

- Diagnose the situation
- Isolate the emergency site from mainline operations
- Implement control and containment measures at the emergency site
- Maintain mainline operations and assure they do not contribute to the escalation of the emergency
- After the emergency is neutralised, reintegrate the affected site into mainline operations.

The optimal emergency control system then is one which provides for rapid assessment of an incident, means and ability to quickly isolate and attack the problem, minimises the effects of the incident on mainline operations and vice versa, and provides for smooth reintegration of the affected site or facility with mainline operations.

It is C&W's experience the more effective emergency plans are found where an organisation has used its 'day to day' control systems (here we are using the term 'controls' to include both procedures and equipment) as its foundations for their emergency plans and procedures supplementing them where necessary. It has been our experience that "bolted on" systems suffer bolt failure (either from metal fatigue or loss of tensile strength).

5 DEVELOPMENT OF A COMMAND SYSTEM

In their investigations into King's Cross Underground Fire and the Piper Alpha Disaster, both Mr Fennell and Lord Cullen found faults in the command of the actual emergency. As with Lord Cullen, Mr Fennell felt that the management had not properly equipped those people on site with the necessary skills and competency to properly manage the events.

It is C&W's experience that breakdown of the planned communication chain occurs very early in major emergencies. This breakdown often occurs while the incident is in a "manageable stage" thus hindering any ability to diagnose the ongoing situation in order to determine how the situation should be tackled. Such breaks contribute to the sequence of escalation either by not being able to isolate the site from mainline operations or by not being able to remove individuals from the danger zone or potential danger zones. While some of these breaks in communications arise from the failure of hardware (monitoring systems, telephones, etc) the human factor has also been found to be particularly frail in this regard.

When developing an emergency plan it is vital to provide a communication system that is sufficiently redundant in order to absorb potential breaks in the system. The communication system must provide concise, sufficient and timely information to:

- those on site or within the potential danger zone
- those upstream and downstream of the affected site or operation
- on and offsite emergency responders
- the media or general public.

It is often the thirst for information of this last group (the media) which causes the communication system to fail as it is insufficient to cope with the demands placed on it by the media or those not directly connected to the event.

Emergencies are exceedingly stressful events. They demand decision-making under the worst of conditions. One is either faced with what seems little accurate information or is swamped in data oft in a format that is difficult to interpret. The emergency plan needs to recognise this and develop a manageable span of control. In some circumstances it may require the span of control be reduced to a one on one situation. In today's environment of "delayering" and "reengineering" to flatter more networked management structures, it might appear traditional but vertical command is desirable in emergency situations to increase supervision and command.

6 AUDITING THE EMERGENCY CAPABILITY

A well constructed audit of the emergency capability of an organisation or of a particular facility will address the systems in place for identifying the hazards and consequences which must be addressed along with the effectiveness of the established plans to control such events, that all required resources are available, maintained and functioning and that the plans are periodically exercised and that any deficiencies found are fedback into the planning process.

The auditor(s) must identify and meet with the key individuals who have the responsibility for emergency response planning as well as those in key positions for implementing the plans and procedures. Audits should verify the following points:

- Systems for identifying an emergency plan(s) is required
- Systems for developing required plans
- Systems for obtaining and maintaining required resources
- Systems for interfacing with outside agencies or support groups
- Systems for testing the plans
- Systems for criticising plans and ensuring changes are implemented
- Systems for assuring plans meet all required legislation

The audit team must be delegated the necessary authority and report to a sufficiently senior level of management to ensure their findings are acted upon by the organisation.

Crisis management and a corporate response organization

M W HOWARD, BSc, CEng, MICE, MIMechE
The British Petroleum Company plc, London, UK

SYNOPSIS A Corporate Crisis Management Plan and Organisation to deal with a major incident from the site to the highest level in The British Petroleum Company

1 INTRODUCTION

The majority of incidents we are involved in attract local or, at the most, national attention. These incidents are dealt with in a relatively short period of time and their effects are minimal.

However, every so often a major crisis event occurs.

If not handled effectively at all levels of the organisation, such an event can seriously damage a Corporation's image, its ability to operate and its finacial well being.

For effective Crisis Management we need to be prepared. we need to establish effective lines of communication both internally and to key areas externally to get our meesage across and to help to mitigate the effects of the crisis.

To help prepare the Organisation for a crisis - by its nature an unexpected and unpredictable event - we need to:-

a. carry out **risk assessment** and **contingency planning** to be aware of the Strategic issues that need to be addressed and their potential impact on the company. To set in place the organisational structure, and identify roles and responsiblities of the people to deal with the issues.

b. carry out regular training and exercises covering a large range of credible scenarios. From these we can:-

1. identify the **strengths and weaknesses** of the Organisation and its people.

2. Test the **organisational procedures** and **lines of communication.**

3. Develop and practice the **roles and responsibilities** for the people and carry out the necessary **training.**

2 CRISIS

It is important to emphasis what is meant by a crisis to understand the reason for the procedures and facilities BP have put in place to manage them.

2.1 Crisis Event

- is a major failure in the company's systems, serious accident or dangerous occurence creating hazards and threats to people, property and/or the environment

- is the sudden and unforeseen event involving extensive damage and loss of control requiring urgent action to restore safe and effecient operations

- could seriously affect the company's image undermine its commitment to social responsibility and its right to operate and affect it's earnings and therefore its financial wellbeing.

In our first plans the term crisis described an event which only affected the BP Group of Companies worldwide.

However in developing the plan it became apparent that issues affecting people, property or the working environment, which needed to be addressed at a Group Corporate level were similar to those issues facing an Associate and/or a

Business at a local level in a particular country or region during an emergency event.

Clearly, this is a matter of perception. From the local community point-of-view, many of the consequences even from a small incident and their local impact may be considered a crisis. Such events need to be dealt with rapidly and sympathetically by the business or local company organisation.

In a major incident there may be national and international consequences which begin to affect the Company's operation in the country, region or the Group as a whole.

To effectively deal with these elements in a crisis a Group Crisis Management Plan and Organisational Structure has been developed combining the principles of Crisis Management with the intricacies of the BP organisation.

3 CORPORATE CRISIS

In a major crisis there is an explosion of impacts which affect the constituents that make up the companies operating environment.

It is the reaction from these constituents which govern the scale of impact that the crisis has on the company and the actions we take to deal with them.

Internal Impacts

Crisis events produce internal stress in the company.

How the response is managed and the demand on resources can affect employees morale and may ultimately affect production.

Due to the demand on resources to deal with a crisis the organisational structure will be under extreme pressure.

Ongoing operations may also be seriously affected if plants have to shut down or products withdrawn.

Shares may also be affected if shareholders concerns about the company's viability/ ongoing operability and it's response to the crisis are not satisfied.

External Impacts

Outside of the company, the crisis can impact or threaten

- People's lives or livelihood. Families and the public at large can be affected due to loss or injury.

- Property damage or destruction.

- And increasingly, in this enlightened world is the perceived threat to or the actual impact on the physical environment.

Effective management of these internal and external constituents and the communication of our actions and decisions is vital to the successful handling and defusing of the crisis.

In 1989 BP reviewed its organisation and its ability to deal with a major crisis event. With committment from top management we have set up a Crisis Management Organisation covering the four major Businesses and the Corporations interests to ensure that BP could effectively deal with any major incident, **from the site, through to the highest level in BP's corporate management.**

4 TIERED RESPONSE

We have developed a tiered response to major incidents. This response includes an **Incident Management Team** and Plans in each Business at local and regional levels, leading up to a **Business Support Team** and management organisation in the Regional and or Business Headquarters and finally to a **Group Crisis Team** in the Corporate Headquarters to address the Group wide strategic issues.

5 KEY PRINCIPLES

The key principles for crisis management are:

- Develop a contingency plan to identify key players and establish lines of communication both within the company and outside.

- Pre-identify a number of people who have the expertise and the authority to make and implement strategic decisions.

- Ensure your team is trained and are fully aware of their roles and responsibilities.

- Identify an area where the crisis management team can debate issues and decide strategy but can also provide enhanced communication and access to information.

- Build up relations with the media and government.

- Last but not least identify criteria for recognising a crisis or threatened crisis. The use of a checklist will help identify criteria.

There is also another very important principle which takes into account how the corporation operates with the Businesses.

The Businesses are totally responsible for the Operational and Tactical response to the emergency, but where issues resulting from the incident affect the Group as a whole, then these issues are addressed by a specially formed group the Group Crisis Team. This team consists of Corporate and Business senior management, who are responsible for the Strategic Management of the crisis.

6 CRISIS MANAGEMENT ORGANISATION

There are three main functions.

- The **Group Crisis Team (GCT)** dealing with the strategic issues affecting the Corporation as a whole worldwide.

  The team is lead by an MD or senior manager appointed by the Chief Executive Officer. Advising him are the heads of Health, Safety and Environment, Human Resources, Corporate Communications, Legal, Finance, Insurance, Regions and the Chief Executive Officer of the affected Business and other Business Chief Executive Officer's as necessary.

- The **Incident Management Team (IMT)** and **Business Support Team (BST)** are teams who respond to the emergency with technical expertise and resources.

  Their function is to respond to and contain the incident, minimise the effect on the Business and ensure the ongoing operation of the business.

- Then there is the **Group Crisis Centre (GCC)** which provides administrative support and information to the Group Crisis Team. In peacetime the Group Crisis Centre is the centre for all crisis management issues in the BP Group, administers the plan and designs and conducts training and exercising of the various teams and procedures.

As mentioned earlier, we have separated crisis management from emergency response and unless there were exceptional circumstances, the Group Crisis Team would not interfere with the Business's response to the incident.

For this reason we have identified specific roles for the Corporation and Business groups

7 PRIMARY ROLE OF THE GCT

- To assess the information and decide whether an incident is stable or escalating into a crisis.

- It must assess the quality of information to ensure its decisions are founded on complete and factual information.

- It has to analyse the information and interpret data and identify Group concerns and strategic issues.

- Determine the company's position with regard to the impact the crisis has on it's

  - image and reputation
  - operability
  - liability and loss potential.

- Determine its strategy for managing constituents considering future implications.

- Review the Business response and its effectiveness

The GCT is supported by a group of people we call the **ADC'S (Aide-de-camp)**,

8 ADC PRIME FUNCTION

The ADC is the representative of the GCT principal. Their role is to research and gather information and assist the principal in analysing and interperating data. There are certain ideal requirements for an ADC:

- They need the appropriate authority to delegate specific actions and to implement decisions made by the GCT within the organisation they represent.

- It is particularly important that those ADC's attached to the Businesses are fully familiar with the emergency response plans.

  This will ensure they understand how their organisation is responding to the incident and will be in a better position to advise the GCT Principal.

- They should all be fully familiar with the people and all organisational aspects of the organisation they represent.

- In familiarising themselves with the response plans they should establish links with the appropriate people whom they would be dealing with in an incident.

- They operate in specially developed Crisis Management facilities and obviously to take full advantage of these facilities, they should be fully familiar with them. Regular training sessions are being implemented for the ADCs.

9 PRIMARY ROLE OF THE BUSINESS

- The Business has the authority and responsibility for containing its own incident.

- It must also ensure that the ongoing business is managing during the emergency.

- It should have a well developed contingency plan which identifies resources and expertise available both within the Business and outside.

- It has a responsibility to ensure complete and factual information is fed to the GCT.

- Should it be necessary, it must implement any strategic decisions made by the GCT.

10 CRISIS MANAGEMENT FACILITIES

As mentioned earlier there is a need for crisis management facilities and an area where both the GCT and the support functions can operate with effective and reliable communications to other parts of the BP organisation and the world at large.

The Group Crisis Team meet in a "Quiet Room" for the team to deal with strategic issues and the ADC's operate from an adjacent Support Room.

There is a one way video and sound link from GCT Room to the Support Room. This link allows the ADC's as a group to be aware of what is happening in the GCT Room and the issues that are being addressed.

Status boards in both rooms will ensure the GCT and support teams are aware of the overall situation.

There is also an information link and a video data system which allows media monitoring, databases, and other information to be transmitted to screens in both rooms.

To ensure that the facilities run smoothly two controllers monitor operations in the GCT Meeting and ADC support room.

These controllers will be fully familiar with:

- The operation of communications hardware.
- The procedures for operating all the support facilities.
- Business Emergency Response plans.
- Activation procedures.
- Information and systems available to the Group Crisis Team and ADCs.

Both rooms will also have a logkeeper to ensure significant decisions and actions are properly recorded and information distributed.

A Facilitator assists with the debating process in the GCT room.

A continuously updated log will be available to the Group Crisis Team and ADC's and shown on large VDU screens at their request.

Administration staff operate the support computers and provide secretarial assistance.

A wide variety of television channels and selected radio stations can be constantly monitored along with satellite stations providing worldwide news coverage from stations such as CNN.

Also a number of databases covering Human and Equipment resources, Logistics, external commercial data and cartographics are available on the support computers to help both the GCT, ADC's and support staff.

11 EXERCISES AND TRAINING

Finally, but by no means least, training and exercising are essential.

We have to date carried out four major exercises involving statutory bodies and the emergency services, role playing journalists, next of kin, government officials,etc., (one such exercise involved over a 1000 people both in and outside the BP organisation). We currently plan to run one major exercise per year involving the organisation at large and if possible the external community and training and limited desk top exercises for the

Group Crisis Management Team and support staff every three months throughout the year.

These exercises allow us to identify possible scenarios, areas of risk and develop appropriate contingency plans.

The lessons learnt and experience gained from exercises and training sessions help to prepare the organisation and identify strengths and weaknesses of both the organisational structure and its people.

They help to test the procedures and to identify suitable people and/or training requirements and establish the roles and responsibilities of all those likely to be involved.

Finally, they allow us to involve the external community, to ensure appropriate and well practised links with emergency services and the community in general are in place, and hopefully to establish and develop the confidence of the general public and media in our organisation and the measures we are taking to protect our operating environment.